D0679844

Multiple Regression and Analysis of Variance

an introduction for computer users in management and economics

GEORGE O. WESOLOWSKY
McMASTER UNIVERSITY
HAMILTON, ONTARIO

A WILEY-INTERSCIENCE PUBLICATION
JOHN WILEY & SONS
NEW YORK • LONDON • SYDNEY • TORONTO

Library of Congress Cataloging in Publication Data

Wesolowsky, George O
Multiple regression and analysis of variance.

"A Wiley-Interscience publication."
Includes index.
1. Regression analysis. 2. Analysis of variance.

I. Title.
QA278.2.W47 519.5′35 76-5884
ISBN 0-471-93373-2

Printed in the United States of America

10 9 8 7 6 5

Preface

THE AIMS OF THE BOOK

Multiple regression is an attempt to model or explain a process that generates data by deriving a certain type of linear relationship from that data. It is a technique widely used in, among others, the fields of economics, biostatistics, social science, engineering, and business. Not only is it useful in itself but it also provides the background for other forms of analysis. In this book multiple regression becomes an approach to simple analysis of variance models. The principles applied in regression, however, are also found in more complex techniques such as discriminant analysis and factor analysis.

The everincreasing emphasis on empirical research in many disciplines, along with the easy availability of prepared computer programs, has led many researchers of minimal statistical background to attempt regression analysis. Unfortunately this technique, although easy to apply, is even easier to misapply.

A great deal of expository material is available on regression and the analysis of variance. This material generally falls into two categories. Much of it is mathematically rigorous and complete, with an emphasis on derivation. In this form it is virtually inaccessible to the reader lacking training in mathematical statistics. At the other extreme are expository works that deal with applications but neglect the statistical details necessary for a full understanding and implementation of the technique. This book is aimed at the gap between these extremes.

When one avoids mathematics in statistics, conclusions and formulas tend to appear as if by magic. To counteract this tendency every attempt has been made to build conceptual constructs, or models, that would make many conclusions seem obvious without formal mathematical proofs. The tools for building these models were restricted to simple algebra, geometry, and large doses of intuition. A statistically unsophisticated reader, on leafing through this book, may be intimidated by the frequency with which Greek-letter-infested formulas occur. Closer inspection will reveal that the same basic formulas recur endlessly, for it is a goal of the book to explain as many things as possible within a single basic framework.

Results and formulas are presented in a form intended for easy understanding and not for efficient calculation; this book is primarily for readers who have access at least to a prepared computer program for multiple regression. The mathematical inelegance (or sloppiness) of some of the notation used will make certain readers wince, but again understandability is the excuse.

Points are belabored that many instructors of mathematical statistics courses would consider trivial. Trivial or not, multiple regression has many pitfalls overflowing with people who have tripped into them. Even researchers of considerable statistical knowledge have immortalized "trivial" blunders in scholarly publications.

The necessary statistical background for full use of this text is a one-semester course at the noncalculus level. Although the review in the first chapter is comprehensive, it is tersely written and not intended as a substitute for a first course in statistics. An elementary fluency in summation notation is assumed.

The book is aimed at a variety of users. It could be used in a one-semester or one-quarter course in regression and the analysis of variance for students in business, economics, or other disciplines. It could also serve as a reference for advanced courses that stress empirical projects. Finally, it could serve as a refresher or reference for researchers whose knowledge of regression remains merely as a vague memory of matrix manipulations and for whom analysis of variance has become a fog of summation signs.

It should be stressed, however, that this book is not at all a complete treatment of regression or the analysis of variance. It is an introduction and a first aid. Many useful topics are excluded because of the need for conciseness, because they require more advanced mathematical explanations, or because they are virtually confined to some specific classes of application.

The topics were chosen to be roughly a common denominator for the generally available computer packages. Neither efficient data-handling techniques nor the comparative virtues of different packages are discussed.

CONTENTS OF THE BOOK

The first chapter is a review of those basic statistical concepts and formulas that are repeatedly used in later chapters. Readers with a recent background in statistics can probably skim over this chapter and refer back to it only when needed.

The second chapter establishes the basic regression model and explains the major part of the terminology appearing on the usual computer printout. It would be dangerous to stop reading at this point.

The third chapter begins with multicollinearity, a condition responsible for much misuse of regression. It then focuses on commonly made errors in interpreting various coefficients and tests. It concludes with a brief discussion of interpolation and extrapolation for purposes of prediction.

The fourth chapter is a discussion of the rationale and criteria for selecting or excluding variables for a regression. Specifically, the well-known techniques of forward selection, stepwise regression, and backward elimination are introduced and illustrated.

Linear regression may be used to establish models that at first appear to be hopelessly nonlinear. In the fifth chapter such techniques as polynomial fitting, transformations, and dummy variables are dealt with.

The sixth chapter in the book deals with some cases in which the actual data-generating process is different from the regression model used to represent it. Consequences on the validity of results are discussed along with remedial measures.

The last two chapters discuss "fixed effects" analysis of variance models. Chapter 7 uses single-factor analysis of variance to develop the regression approach that is applied to the widely used experimental designs in Chapter 8. The regression approach has some advantages and disadvantages over the traditional exposition of analysis of variance. Once multiple regression is mastered analysis of variance models can be explained without recourse to new computational formulas. "Traditional" computer printouts can easily be understood and used within this framework. In the case of "balanced" data the actual use of regression is cumbersome compared with traditional computational procedures. When data is unbalanced because of missing or invalidated observations (a common occurrence in the social sciences), the actual use of regression computation for experimental designs comes into its own as a powerful tool.

FEATURES OF THE BOOK

There are exercises at the end of every chapter of the book except the first (a solutions manual is available to instructors from me). When these exercises present data for analysis, they tend to be somewhat artificial. If this book is used in a course, the instructor should have no difficulty in providing data relevant to the discipline of interest. Access to a computer and prepared computer packages is assumed. The examples and some of the exercises make use of tables derived from computer outputs. When possible, the original (and often peculiar) labeling, headings, and format have been preserved. This was done to encourage the development of deciphering skills and to provide familiarity with different terminologies.

Two appendices are provided. The "Mathematical Notes" appendix contains some of the simpler proofs. It should enable students with a degree of mathematical sophistication to feel more "comfortable" about the results given. Some of the notes do require a knowledge of elementary calculus. The book is intended, however, to be understandable without them.

An appendix on matrices is an introduction to the terminology used;

matrix terminology often appears on computer printouts. This appendix, however, gives sufficient matrix algebra for the understanding of the one mathematical note that uses matrix manipulations.

GEORGE O. WESOLOWSKY

Hamilton, Ontario
January 1976

Acknowledgments

In common with the practice of most other authors, I have thoroughly exploited many friends and colleagues during the preparation of this manuscript. Professors Anne Sarndal, George W. Torrance, and William G. Truscott deserve gratitude for suffering through the major part of the book and for their comments and suggestions. Professors Kenneth R. Deal and Bent Stidsen unflinchingly read the first third of the book.

My special thanks go to Professor Bill Richardson who pounced on many, many errors and to Professor Charles H. Goldsmith who helped with the last two chapters. I am also indebted to my students who hunted errors at the niggardly pay of a penny apiece and to a nameless referee. Any errors that remain should reasonably be attributed to those that failed me in detecting them. However, I will bow to the hoary weight of tradition and accept the blame myself.

Mrs. Barbara Dove typed superbly in turn each of the "final" versions that I gave her. Sharon Wesolowsky, a relative by marriage, also exhibited great patience: before the writing of this book, during the writing, and (hopefully) afterward.

G.O.W.

Contents

chapter 1

A review of statistical concepts

1.1 RANDOM VARIABLES

A *random variable* is a function or rule that assigns a number to each possible outcome of an experiment or sample; for example, when a coin is tossed three times, the number of heads that occurs can be considered a random variable. This random variable can take the values 0, 1, 2, or 3.

Random variables are either discrete or continuous. *Discrete random variables* can take only isolated values; the number of heads that occurs in three tosses of a coin is a discrete random variable. A *continuous random variable*, on the other hand, can take any value within a range of values. The weight of an adult mouse chosen from a population of mice can take any value within the range of weights for adult mice. Continuous random variables are usually involved when the data consist of measurements rather than counts.

1.2 PROBABILITY DISTRIBUTIONS

When a measure of probability is assigned to each possible outcome for a random variable X, a *probability distribution* is produced. A *discrete probability distribution* is simply a list of probabilities, one of which is assigned to each possible value of a random variable. The sum of the probabilities must always equal 1.

▶ **Example 1.1** If X is the total number of heads in three tosses of a coin, the probability distribution for X can be given by[1] Table 1.1 or by the function

$$\Pr(X) = \frac{3!}{X!\,(3-X)!} \cdot \left(\tfrac{1}{2}\right)^{X} \left(\tfrac{1}{2}\right)^{3-X} \qquad \text{for } X = 0, 1, 2, 3.$$

[1]This result comes by way of the binomial distribution, a discrete probability distribution described in introductory statistics texts.

Table 1.1 Probability Distribution for the Number of Heads in Three Tosses of a Coin

X	0	1	2	3
Pr (X) (probability of X)	$\frac{1}{8}$	$\frac{3}{8}$	$\frac{3}{8}$	$\frac{1}{8}$

◀

It is not possible to provide a probability list in tabular form for a continuous random variable. For one thing, a random variable, which can take any of an infinite number of possible values, would need a rather long table. For another, every individual probability value within that impossible table would be zero, for example, the probability that a mouse picked at random will weigh *exactly* 1 ounce, to an infinite number of zeros after the decimal point, is 0.

The probability distribution for a continuous random variable is given by what is called a *probability density function*. If the probability density function is plotted against X, the area under its curve (bounded by the X axis) is equal to 1. Furthermore, the area under the curve between any two X values $X = a$ and $X = b$ is equal to the probability that X will lie between a and b.

▶ **Example 1.2** Consider a computer that is generating random numbers X between .00000000 and .99999999. Each number in the range is equally likely to appear. We could consider the random number X to be an approximately continuous random variable with a range 0 to 1.

Because no possible value of X is any more probable than any other, the probability density function must have the same value 0 to 1. This is illustrated in Figure 1.1. The area under the curve must equal 1, and because the width of the rectangle is 1 the height therefore must also be 1.

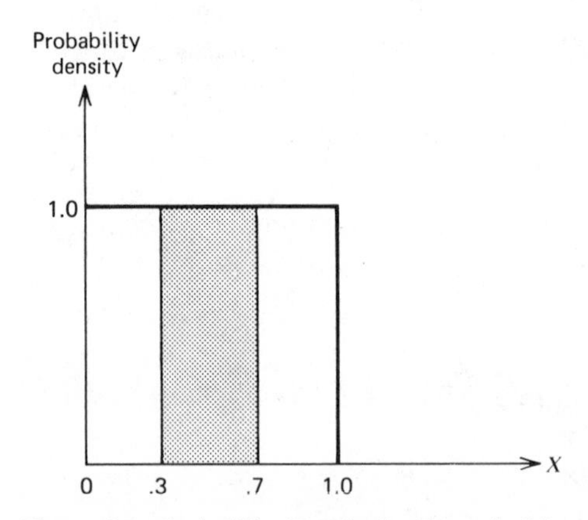

Figure 1.1 Probability Density for Example 1.2

Suppose that we wanted to find $\Pr(.3 \leqslant X \leqslant .7)$. In this case, $a = .3$ and $b = .7$. Therefore $\Pr(.3 < X < .7) = (.7 - .3)(1) = 0.4$ ◀

1.3 EXPECTED VALUES

The expected value of a variable can be thought of loosely as a long-run average of that variable. More formally, for a discrete variable X the *expected value* $E(X)$ is given by

$$E(X) = \sum_{i=1}^{n} X_i \Pr(X_i) \tag{1.1}$$

The subscript i serves to identify each of the n possible values of X.

▶ **Example 1.3** What is the long-run average number of heads we may expect when we keep repeating the experiment of tossing a coin three times? We expect that there will be no heads about $\frac{1}{8}$ of the time, one head about $\frac{3}{8}$ of the time, two heads about $\frac{3}{8}$ of the time, and three heads about $\frac{1}{8}$ of the time. The expected value, or long-run average, of the number of heads is therefore

$$E(X) = 0(\tfrac{1}{8}) + 1(\tfrac{3}{8}) + 2(\tfrac{3}{8}) + 3(\tfrac{1}{8}) = \frac{12}{8} = 1\tfrac{1}{2}$$ ◀

The expected value of a continuous random variable is

$$E(X) = \int_{r_1}^{r_2} X f(X)\, dX \tag{1.2}$$

where the range of X is r_1 to r_2 and $f(X)$ is the probability density function of X. An integral[1] and not a summation is used because a continuous random variable can take an infinite number of values in its range.

The following rules are useful in operating with expected values:

1. $E(C) = C$. The expected value of a constant C is the constant itself.
2. $E(CX) = CE(X)$. The expected value of a constant times a variable is the constant times the expected value of the variable.
3. $E(X \pm Y) = E(X) + E(Y)$. The expected value of a sum (or difference) of two variables is the sum (or difference) of the expected values of those variables.

[1]The reader should refer to Note 1.1 in Appendix A for an explanation of the integral if his calculus is somewhat rusty.

For instance, if X is the number of heads in three tosses of a coin, then

$$E(3X + 2) = E(3X) + 2 = 3E(X) + 2 = 6.5$$

1.4 VARIANCE

The *variance of a random variable*, whether discrete or continuous, is the expected value of the squared deviation of the variable from the expected value of that variable:

$$\mathrm{VAR}(X) = E\{[X - E(X)]^2\} \tag{1.3}$$

▶ **Example 1.4** If X is the number of heads obtained in tossing a coin three times, $E(X) = 1.5$. Table 1.2 gives the probability distribution for $[X - E(X)]^2$. The expected value of $[X - E(X)]^2$ is

$$(-1.5)^2(\tfrac{1}{8}) + (-.5)^2(\tfrac{3}{8}) + (.5)^2(\tfrac{3}{8}) + (1.5)^2(\tfrac{1}{8}) = .75$$ ◀

Table 1.2 Probability Distribution for $(X - E(X))^2$, where X is the Number of Heads in Three Tosses of a Coin

X	0	1	2	3
$[X - E(X)]^2$	$(-1.5)^2$	$(-.5)^2$	$(.5)^2$	$(1.5)^2$
$\Pr([X - E(X)]^2) = \Pr(X)$	$\frac{1}{8}$	$\frac{3}{8}$	$\frac{3}{8}$	$\frac{1}{8}$

The expected value of a random variable (sometimes called the mean of the random variable) is one that positions the "center" of its probability distribution on the X axis. The variance of a random variable is a measure of the dispersion of that probability distribution; for example, in Figure 1.2 probability distribution 2 will have a greater variance because it is more "spread out" on the X axis than probability distribution 1.

In addition to measuring the dispersion of a probability distribution, we also apply the term variance in descriptive statistics to a population of N values or observations. A measure of the dispersion of the population along the X axis is

$$\mathrm{VAR}(X) = \frac{\sum_{i=1}^{N} (X_i - \mu_X)^2}{N} \tag{1.4}$$

where $X_i = 1, \ldots, N$ are the observations and $\mu_X = \sum_{i=1}^{N} X_i / N$ is the average value in the population. The symbol μ_X is also often used to depict $E(X)$. The population of numbers {1, 2, 3} would have $\mu_X = (1 + 2 + 3)/3 = 2$ and variance $(1^2 + 0^2 + 1^2)/3 = \frac{2}{3}$. The *standard deviation* is always the positive square root of variance and is written as $SD(X)$.

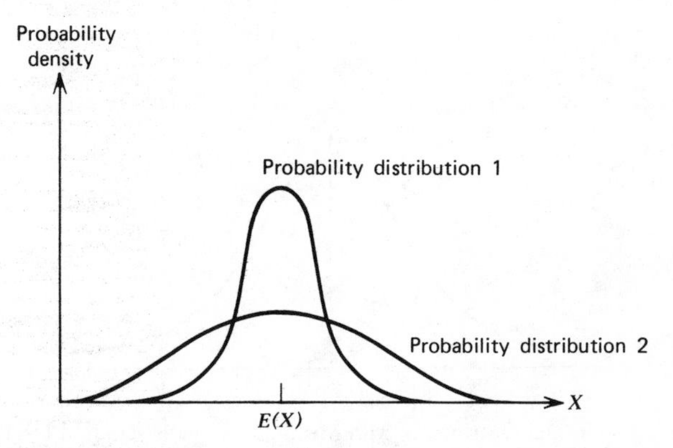

Figure 1.2 Comparison of the Variances of Two Distributions

As with expected values, we can perform operations with variances:

1. $VAR(X + C) = VAR(X)$. The variance of a random variable plus a constant is equal to the variance of that random variable.
2. $VAR(XC) = C^2VAR(X)$. The variance of a random variable multiplied by a constant is the constant squared times the variance of the variable.
3. $VAR(X + Y) = VAR(X) + VAR(Y) + 2E\{[X - E(X)][Y - E(Y)]\}$. The variance of the sum of two random variables X and Y is the sum of the variances of X and Y plus twice $E\{[X - E(X)][Y - E(Y)]\}$, a quantity called the *covariance* of X and Y. More is said about covariance in Section 1.13.

1.5 THE NORMAL DISTRIBUTION

A continuous random variable X, with a possible range of $-\infty$ to $+\infty$ and with an expected value μ_X and standard deviation σ_X has a *normal probability* distribution if its probability density is given by

$$f(X) = \frac{1}{\sqrt{2\pi}\,\sigma_X} \exp\left[-\frac{1}{2}\left(\frac{X - \mu_X}{\sigma_X}\right)^2\right] \tag{1.5}$$

When we plot $f(X)$ against the X variable, we obtain the bell-shaped curve in Figure 1.3. The curve is symmetrical about the value μ_X and the area under the curve is 1. By using the definition of expected value in (1.2) we can verify that $E(X) = \mu_X$ and $E[(X - \mu_X)^2] = \sigma_X^2$.

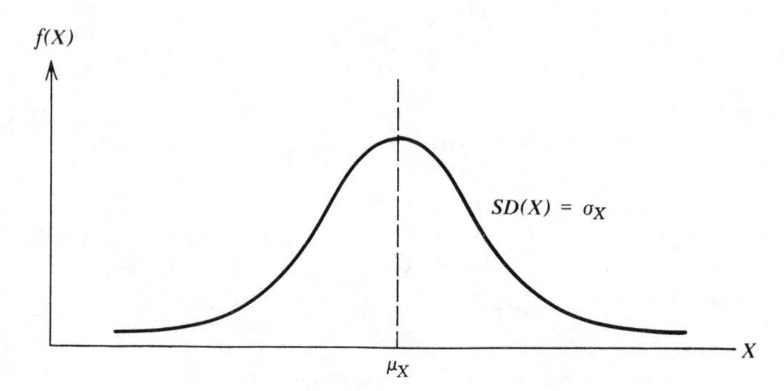

Figure 1.3 Normal Distribution with Mean μ_X and with Standard Deviation σ_X

The normal distribution is central to the mathematical model used in multiple linear regression. It is, as well, the most widely applied distribution in statistics. It would be useful to have tables to read such probabilities as $\Pr(C_1 \leqslant X \leqslant C_2)$, where C_1 and C_2 are some constants. Unfortunately this would mean that we would need to construct tables for every possible value of μ_X and σ_X. Fortunately, by using a rather minor trick it is always possible to find such a probability by referring to a table for a normal distribution with mean 0 and standard deviation 1.

1.6 THE STANDARDIZED NORMAL DISTRIBUTION

Consider a normal distribution of a random variable X with an expected value μ_X and standard deviation σ_X. We can define a new random variable z where

$$z = \frac{X - \mu_X}{\sigma_X} \tag{1.6}$$

Because μ_X and σ_X are constants, z also has a normal distribution; for every value of X there is a corresponding value of z.

The expected value of z is

$$
\begin{aligned}
E(z) &= E\left(\frac{X - \mu_X}{\sigma_X}\right) \\
&= \frac{1}{\sigma_X} E(X - \mu_X) && \text{Rule 2, Section 1.3} \\
&= \frac{1}{\sigma_X}(E(X) - \mu_X) && \text{Rules 1 and 3, Section 1.3} \\
&= 0
\end{aligned}
$$

The variance of z is

$$
\begin{aligned}
\text{VAR}(z) &= E(z - E(z))^2 \\
&= E\left(\frac{X - \mu_X}{\sigma_X} - 0\right)^2 \\
&= \frac{1}{\sigma_X^2} E(X - \mu_X)^2 && \text{Rule 2, Section 1.4} \\
&= \frac{1}{\sigma_X^2} \cdot \sigma_X^2 \\
&= 1
\end{aligned}
$$

The random variable z has mean 0 and standard deviation 1 and is said to have a *standardized normal distribution*. Probability tables for z are readily available and can be utilized to find probabilities for any normally distributed random variable.

▶ **Example 1.5** Find $\Pr(30 \leqslant X \leqslant 60)$ if X is normally distributed with mean 50 and standard deviation 20 [$N(50, 20)$ is a short way of describing the distribution of X]. The probability wanted is equal to the shaded area between $X = 30$ and $X = 60$ in Figure 1.4. Because there is direct correspondence between X and z, this area is equal to the area between $z = z_1$ and $z = z_2$.

From (1.6) $z_1 = (30 - 50)/20 = -1$ and $z_2 = (60 - 50)/20 = \frac{1}{2}$. Table 1 in Appendix C gives the area to the left of any z. $\Pr(z \leqslant \frac{1}{2}) = .6915$ and $\Pr(z \leqslant -1) = .1587$. Therefore $\Pr(z_1 \leqslant z \leqslant z_2) = \Pr(30 \leqslant X \leqslant 60) = .6915 - .1587 = .5328$. ◀

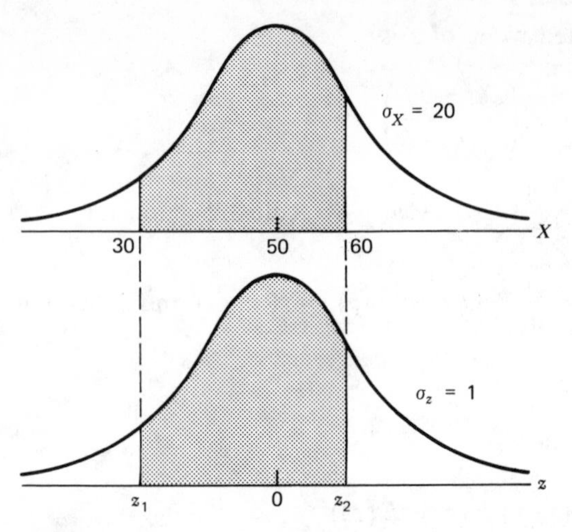

Figure 1.4 The X and z Distributions in Example 1.5

1.7 SAMPLING DISTRIBUTION OF THE MEAN

In our discussion of multiple regression we make extensive use of random variables called statistics. A *statistic* is a random variable whose values are calculated from sample data. It is a random variable because the same calculation for different samples from the same population may produce different values. The variance of this random variable depends on the size of the population, the variance of the population, the size of the sample, and, in the case of a finite population, whether we sample with or without replacement; for example, $\overline{X} = \Sigma_{i=1}^{n} X_i/n$, called the *sample mean*, is a statistic. It is the average of n values chosen from a population in a manner that gave each value the same chance of being chosen. The statistic $\overline{X}$ is often used in estimating a population mean or in testing hypotheses concerning a population mean. Technically, of course, a single value X randomly chosen from a population is itself a statistic.

A statistic, being a random variable, will have a probability distribution, which is generally called a *sampling distribution*; its standard deviation is the *standard error* of the statistic.

The distribution of $\overline{X}$ is described by two theorems; one gives the expected value and the variance and the other, the shape of the distribution.

Theorem.[1] If an infinitely large population has a mean μ_X and standard deviation σ_X, then $E(\overline{X}) = \mu_X$ and $\text{VAR}(\overline{X}) = \sigma_{\overline{X}}^2 = \sigma_X^2/n$.

[1]For a proof see Appendix A, Note 1.2.

If sampling were done from a finite population, VAR($\overline{X}$) in this theorem would have to be modified. In this book we deal with the theoretically infinite populations.

Central Limit Theorem. If an infinitely large population has a mean μ_X and standard deviation σ_X, the distribution of the sample mean $\overline{X}$ approaches the normal distribution with mean μ_X and standard deviation $\sigma_{\overline{X}} = \sigma_X / \sqrt{n}$ as the sample size n increases.

Generally (except for very unusual population distributions) a sample of size 30 is large enough for $\overline{X}$ to be distributed with a nearly normal distribution. If the population distribution (denoted by the probability distribution of X) is normal, the distribution of $\overline{X}$ is exactly normal, regardless of the size of n. When $\overline{X}$ has a normal distribution, then

$$z = \frac{\overline{X} - \mu_X}{\sigma_{\overline{X}}} \tag{1.7}$$

has a standardized normal distribution and can be used to provide probabilities.

▶ **Example 1.6** Suppose that we are sampling weights of individuals from a large urban population with a known average weight of 140 pounds and a known standard deviation of 30 pounds. If our samples are size 100, then $\overline{X}$, the sample mean, will have an approximately normal distribution with mean $E(\overline{X}) = 140$ and standard deviation $\sigma_{\overline{X}} = 30/\sqrt{100} = 3$. Suppose that we wish to find $\Pr(\overline{X} \geqslant 141)$, the shaded area in Figure 1.5.

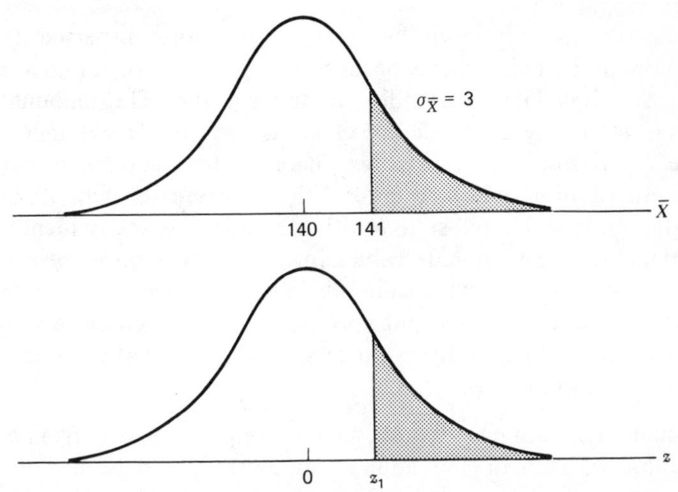

Figure 1.5 Probability Distribution of Sample Means and the Appropriate z Distribution in Example 1.6

Then $\Pr(\bar{X} \geqslant 141) = \Pr(z \geqslant z_1)$, where $z_1 = (141 - 140)/3 = \frac{1}{3}$. Hence $\Pr(\bar{X} \geqslant 141) = 1 - .6293 = .3707$ (from Table 1 in Appendix C). ◀

1.8 THE *t* DISTRIBUTION

The t distribution is useful in regression analysis. Consider first the distribution of $z = (\bar{X} - \mu_X)/\sigma_{\bar{X}}$. It has a standardized normal distribution because μ_X and $\sigma_{\bar{X}}$ are constants. Suppose σ_X is not known but has to be estimated from the sample standard deviation. In other words, we use

$$S_X = \left[\frac{\sum_{i=1}^{n} (X_i - \bar{X})^2}{n - 1} \right]^{1/2} \tag{1.8}$$

to estimate σ_X. Equation 1.8, when both sides are squared, is a direct analog of (1.4) except for the 1 subtracted from n. The reason for this difference is discussed in Section 1.10. Because S_X, as well as $\bar{X}$, is a random variable, we use a different symbol t for the statistic $(\bar{X} - \mu)/(S_X/\sqrt{n})$:

$$t = \frac{\bar{X} - \mu_X}{S_X/\sqrt{n}} = \frac{\bar{X} - \mu_X}{S_{\bar{X}}} \tag{1.9}$$

where $S_{\bar{X}} = S_X/\sqrt{n}$.

The distribution of t will be "flatter" or more dispersed than the distribution of z. This is true because the presence of another random variable S_X gives more variability to the statistic. The amount of this additional variability is a function of n. As the sample size increases the estimate S_X becomes more stable. There is, for small n, a noticeably different distribution of t for every n. When n is greater than 30, however, S_X is sufficiently stable (close to σ_X) that z and t are nearly identical.

One point should be noted. Tables for t are not in terms of n but of ν ($\nu = n - 1$), a quantity that is called the *degrees of freedom* for the statistic $(\bar{X} - \mu_X)/S_{\bar{X}}$. Much more is said about degrees of freedom in Chapter 2. Tables for t generally go only as far as $\nu = 30$; after that the standardized normal distribution is used.

▶ **Example 1.7** Suppose that we take a sample of size 30 from a normal distribution with mean 140 and an unknown standard deviation. The statistic $(\bar{X} - 140)/(S_X/\sqrt{30})$ will have a t distribution. Reading Table 2 in Appendix C at 29 degrees of freedom, we find that the probability that this statistic will have a value greater than 2.462 is .01. ◀

Other statistics also have the t distribution, although for them ν is not always equal to $n - 1$. Usually we have to remember to use the t table instead of the standardized normal table if the standard deviation of the statistic is not exactly known but only estimated. It should be noted, however, that the t table is generally less complete than the normal table because it is used mainly for estimation and hypothesis testing. In those applications only certain values of probability are traditional.

1.9 THE *F* DISTRIBUTION

Another distribution that describes many useful statistics is the F distribution. Consider, for example, the statistic S_Y^2/S_X^2, which is calculated from data obtained by taking a sample of size n_Y from the normal distribution for Y and another of size n_X from the normal distribution of X; S_Y^2 and S_X^2 are calculated from Equation 1.8.

If it is true that $\sigma_Y^2 = \sigma_X^2$, the statistic S_Y^2/S_X^2 has a distribution shown in Figure 1.6. Like the t distribution, the shape of this distribution, which is called an *F distribution*, is a function of sample sizes. Naturally there are no negative values of F.

Statistics having the F distribution are ratios of two other statistics. The F distribution therefore has two values for degrees of freedom: one referring to the numerator and one to the denominator. For S_Y^2/S_X^2, $\nu_1 = n_Y - 1$ and $\nu_2 = n_X - 1$.

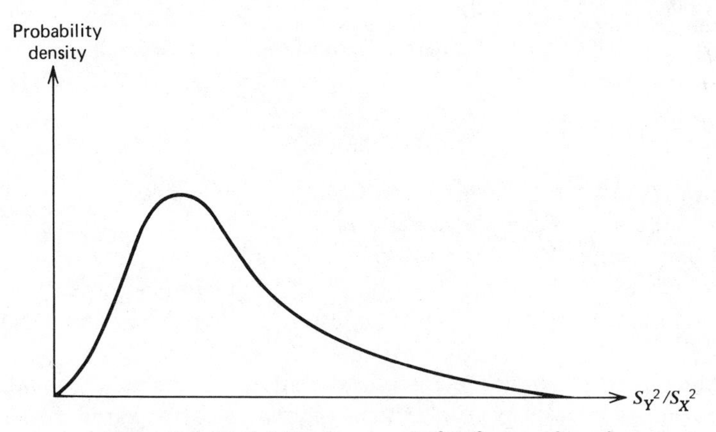

Figure 1.6 Probability Density of S_Y^2/S_X^2 when $\sigma_Y^2 = \sigma_X^2$

▶ **Example 1.8** There are two normal populations with the same variance. We take a sample of $n_Y = 31$ from the first and $n_X = 17$ from the second. What is the probability that the ratio of sample variances S_Y^2/S_X^2 is greater than 2.20?

From Table 3 in Appendix C, using $\nu_1 = 30$ and $\nu_2 = 16$, we obtain $\Pr(F > 2.20) = .05$. Naturally, if a value other than 2.20 had been requested (unless it were 3.1), we would have had to consult other tables. As with t tables, F tables are usually given in limited form.

It is interesting (as well as important) to note at this point that the statistic

$$t^2 = \left[\frac{\bar{X} - \mu_X}{S_X/\sqrt{n}} \right]^2 \tag{1.10}$$

has an F distribution with $\nu_1 = 1$ and $\nu_2 = n - 1$; for example, from Table 2 at $\nu = 15$

$\Pr(-2.131 \geqslant t \geqslant 2.131) = .05 = \Pr(t^2 \geqslant (2.131)^2) = \Pr(t^2 \geqslant 4.54)$
From Table 3 $\Pr(F \geqslant 4.54) = .05$ at $\nu_1 = 1$ and $\nu_2 = 15$. This is illustrated in Figure 1.7. ◀

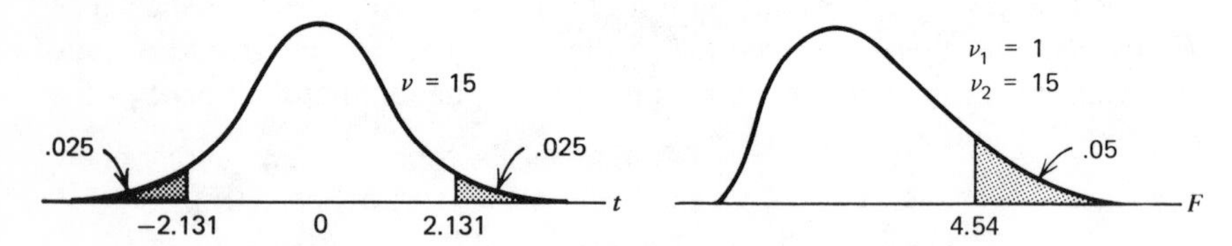

Figure 1.7 The t and F Distributions in Example 1.8

1.10 ESTIMATION

We have reviewed the distribution of some statistics—random variables—that are functions of sample values. Many others, for example, $\sin(\Sigma_{i=1}^n x_i)$, would be possible. We wish, however, to use only those that are best suited to our purposes: the estimation of population *parameters* such as the population mean or variance and the testing of hypotheses about parameters.

Turning first to the problem of estimation, we judge the quality of statistics as estimators by the criteria of bias, mean squared error, consistency, and maximum likelihood.

Bias

An estimator $\hat{\theta}$ is said to be an unbiased estimator for a parameter θ if $E(\hat{\theta}) = \theta$, that is, if the "long-run" average of the random variable $\hat{\theta}$ is θ, the value of the parameter. For example, $\bar{X}$ is an unbiased estimator of μ_X, the population mean, because $E(\bar{X}) = \mu_X$. Also, S_X^2 in (1.8) is an unbiased estimator for σ_X^2, the population variance. The divisor $n - 1$ was used in (1.8) because the use of n would have resulted in a biased estimator. It should be noted that S_X is not an unbiased estimator for σ_X. To understand this consider the analogous example of the estimator $\hat{\theta}$ in Table 1.3. It is easy to show that

$$E(\hat{\theta}) = 2$$

$$E(\hat{\theta}^2) = 4.67$$

If $\hat{\theta}^2$ is an unbiased estimator for $\theta^2 = 4.67$, then $\hat{\theta}$ is obviously not an unbiased estimator for $\theta = 2.16$.

Lack of bias is a desirable, but not necessarily indispensible, property of an estimator.

Table 1.3 Probability Distributions for $\hat{\theta}$ and $\hat{\theta}^2$

$\hat{\theta}$	1	2	3
$\Pr(\hat{\theta})$	1/3	1/3	1/3
$\hat{\theta}^2$	1	4	9

Mean Squared Error

The mean squared error of an estimator $\hat{\theta}$ is $E(\hat{\theta} - \theta)^2$, or the expected value of its squared deviation from the parameter θ. For unbiased estimators the mean squared error is obviously equal to the variance of the estimator. For example, we can calculate $\bar{X}$ or the median $\hat{M}$ from a sample of 100 drawn from a normal population. The median is the middle value in a sample with data arranged by size if the sample has an odd size n; if n is even, the median is the average of the two middle values. As shown in Figure 1.8, the distribution of $\hat{M}$ is more spread out than that of $\bar{X}$. Hence $E(\bar{X} - \mu_x)^2 < E(\hat{M} - \mu_x)^2$. Because $E(\hat{M})$ can be shown to equal μ_x, $\hat{M}$ is an unbiased estimator for μ_x and $E(\hat{M} - \mu_x)^2$ is the variance of $\hat{M}$.

An unbiased estimator with a smaller variance than another unbiased estimator is said to be *relatively more efficient*. According to this definition, $\bar{X}$ is more efficient than $\hat{M}$.

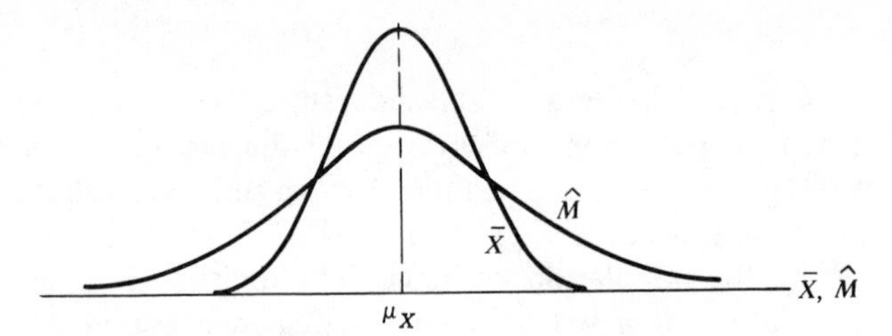

Figure 1.8 Comparison of Distributions for the Sample Mean and the Sample Median

Let us now look at the trade-off between bias and mean squared error. In Figure 1.9 the estimator $\hat{\theta}_2$, although biased, is probably more desirable than the unbiased estimator $\hat{\theta}_1$ because it has a smaller mean squared error and is more likely to produce "close" estimates for θ.

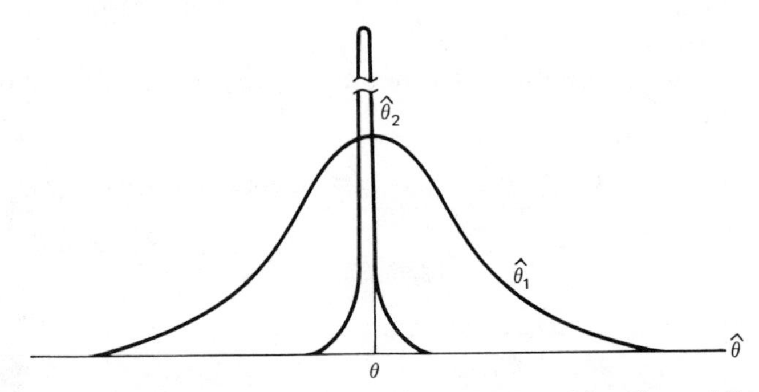

Figure 1.9 Biased and Unbiased Estimators of Different Mean Squared Error

Consistency

An estimator is said to be *consistent* if values of $\hat{\theta}$ can be made to converge to θ by increasing the sample size n; for example, $(\bar{X} + 1/n^2)$ is a consistent estimator of μ_X because $(\bar{X} + 1/n^2) \to \mu_X$ as $n \to \infty$. However, $(\bar{X} + 1/n^2)$ is a biased estimator for μ_X because $E(\bar{X} + 1/n^2) = \mu_X + 1/n^2 \neq \mu_X$.

An unbiased estimator may or may not be consistent. In general, consistency is a desirable property because it gives us hope that larger samples will give better estimates.

Maximum Likelihood

Given a particular set of sample observations, we choose as our estimator of θ the estimator $\hat{\theta}$ that maximizes the probability (or probability density)

of obtaining those observations. The following example should clarify the concept:

▶ **Example 1.9** Suppose that we wish to estimate the probability of heads, $\Pr(H)$, for a biased coin from three flips of that coin.

Let X_i be the number of heads on the ith toss. The variable X_i can take values 0 or 1. Therefore $X_1 + X_2 + X_3$ or $\Sigma_{i=1}^{3} X_i$ is the number of heads in three tosses. The probability of any particular result (X_1, X_2, X_3) for an assumed $\Pr(H)$, or P, is

$$\Pr(X_1, X_2, X_3) = P^{\Sigma X_i} \cdot (1 - P)^{3 - \Sigma X_i} \tag{1.11}$$

For example, $\Pr(1, 0, 1) = P(1 - P)P = P^2(1 - P)^1$. The same result is obtained from (1.11) with $\Sigma X_i = 2$.

The value of P that maximizes (1.11) is our estimator for $\Pr(H)$. By using calculus, we obtain $\hat{P} = \frac{1}{3}\Sigma X_i$; this is our maximum likelihood estimator. ◀

It turns out that the familiar estimator $\bar{X}$ is a maximum likelihood estimator for μ_X. However, S_X^2 is not a maximum likelihood estimator for σ_X^2 [but $(n - 1)S_X^2/n$ is]. The maximum likelihood principle is a good method for searching for estimators. Maximum likelihood estimators have some desirable properties that are beyond the scope of this book to discuss.

1.11 CONFIDENCE INTERVALS

If we obtain an estimate for a parameter, the question is, how good is the estimate? We can answer indirectly by quoting such properties of the estimator as unbiasedness and consistency, but it is usual to approach the problem in another way.

We narrow our interest to unbiased estimators with a normal sampling distribution. Suppose such an estimator $\hat{\theta}$ with standard deviation $\sigma_{\hat{\theta}}$ and expected value $E(\hat{\theta}) = \theta$ is used to obtain an estimate for θ. Let us consider a new statistic in the form of an interval:

$$\hat{I}_{1-\alpha} = \left[\hat{\theta} - z_{\alpha/2}\sigma_{\hat{\theta}}, \hat{\theta} + z_{\alpha/2}\sigma_{\hat{\theta}}\right] \tag{1.12}$$

Let $z_{\alpha/2}$ be a particular value of z from a standardized normal distribution; both $z_{\alpha/2}$ and $\sigma_{\hat{\theta}}$ are constants. The interval $\hat{I}_{1-\alpha}$, however, is a statistic and will carve out different sections of the $\hat{\theta}$ axis each time we sample. The width of the interval is $2 \cdot z_{\alpha/2} \cdot \sigma_{\hat{\theta}}$, and it is centered at $\hat{\theta}$. Figure 1.10 shows three outcomes of a sampling process. Since $\hat{\theta}$ is

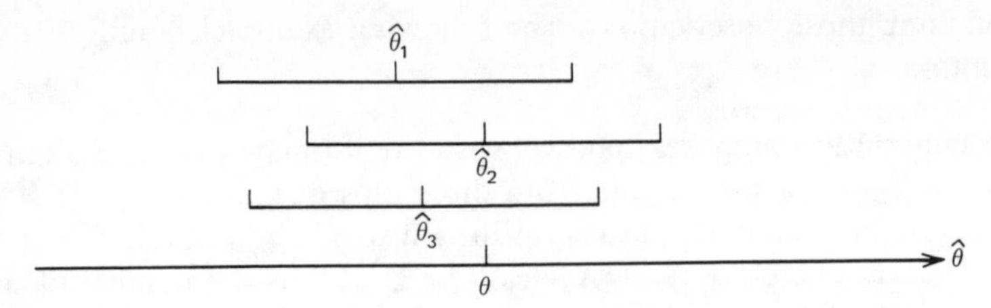

Figure 1.10 Three intervals.

unbiased, we would hope that if the interval width were wide with respect to $\sigma_{\hat{\theta}}$ many of the intervals would "encompass" the true value θ.

If $z_{\alpha/2} = 1.96$, then 95% of the intervals encompass θ, and if $z_{\alpha/2} = 1.645$ then 90% of the intervals encompass θ. The above statements are part of a more general rule proved in most elementary statistics books.

Rule. If $z_{\alpha/2}$ is that value of z such that $\alpha/2 \cdot 100$ is the percent of the area to the right of it, then $(1 - \alpha) \cdot 100\%$ of the intervals will encompass θ.

It is more common to say that (1.12) is the interval at the $(1 - \alpha) \cdot 100\%$ *confidence level.*

▶ **Example 1.10** Suppose that a sample of 64 items is taken from a population with a known standard deviation $\sigma_X = 96.0$. The sample mean $\overline{X}$ is 164.0. Although we use $\overline{X} = 164.0$ as our "point" estimate of the mean μ, we need a measure of the quality of the estimate.

Let us solve this problem by calculating an interval wide enough that 99% of these intervals will encompass the mean μ. Because $1 - \alpha = .99$, $\alpha = .01$ and $\alpha/2 = .005$. Using Table 1 in Appendix C, we have $z_{\alpha/2} = 2.576$ as in Figure 1.11.

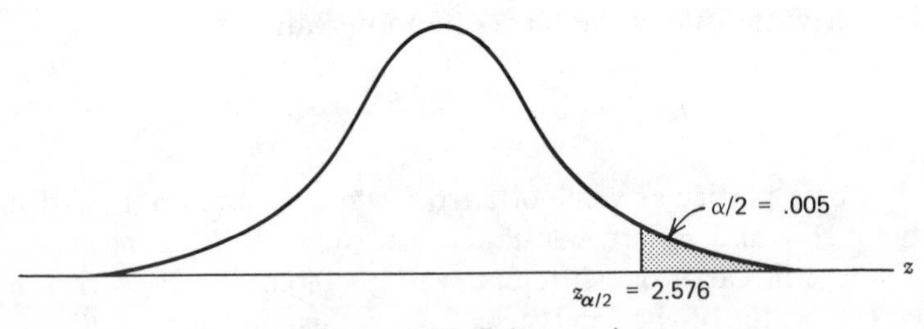

Figure 1.11 $\Pr(z > z_{\alpha/2})$

From the first theorem in Section 1.7 $\sigma_{\bar{X}} = 96/\sqrt{64} = 12.0$. Therefore, from (1.12),

$$\hat{I}_{.99} = [164 - 2.576(12), 164 + 2.576(12)]$$

$$= [133.088, 194.912]$$

This interval is stated at the 99% confidence level. Naturally, if we repeated our calculations at the 95% confidence level, the resulting interval would have a smaller width. It is important to note that our estimator was only approximately normal by the central limit theorem; $\bar{X}$ is exactly normal only if the parent population is normal. If the sample size were not large enough for the central limit theorem to operate, our results would be in error. ◀

There is one further difficulty. Often we do not know the standard error (deviation) of the estimator that we wish to use but must obtain an estimate from the sample. In this case a t distribution with the appropriate degrees of freedom is used and (1.12) becomes

$$\hat{I}_{1-\alpha} = [\hat{\theta} - t_{\alpha/2}S_{\hat{\theta}}, \hat{\theta} + t_{\alpha/2}S_{\hat{\theta}}] \tag{1.13}$$

where $S_{\hat{\theta}}$ is an estimator for $\sigma_{\hat{\theta}}$.

▶ **Example 1.11** Consider a sample of 16 items from a normal population with an unknown standard deviation. The sample mean $\bar{X}$ is 20 and the sample standard deviation S_X is 15. The 99% confidence interval is required.

Since $\hat{\theta} = \bar{X}$, $(\bar{X} - \mu_X)/S_{\bar{X}}$ has a t distribution with $n - 1$ degrees of freedom. From Table 2 in Appendix C, $t_{\alpha/2} = t_{.005} = 2.947$. Further,

$$S_{\hat{\theta}} = S_{\bar{X}} = \frac{15}{\sqrt{16}} = 3.75$$

Now

$$\hat{I}_{.99} = [20 - 3.75(2.947), 20 + 3.75(2.947)]$$

$$= [8.949, 31.05]$$

at the 99% confidence level. ◀

1.12 HYPOTHESIS TESTING

Hypothesis testing is simple in concept. We make an assumption about a population parameter and then calculate a statistic from a sample of the population. If the statistic value is unlikely, given the assumption, we reject the assumption.

The *null hypothesis* is the assumption being tested. We wish to know if there are grounds for rejecting it. The *alternative hypothesis* is an assumption that we offer as an alternative to the null hypothesis. The *critical region* is defined as the set of those statistic values that would lead to the rejection of the null hypothesis. Because the statistic is a random variable, it is possible that the statistic value will fall into the critical region even though the null hypothesis about a parameter is true. It is also possible that the statistic will not fall into the critical region when the null hypothesis is false. Both types of outcome would lead us into making a wrong decision. These wrong decisions are called *Type I* and *Type II* errors, as summarized in Figure 1.12; H_0 stands for the null hypothesis and H_1 for the alternative hypothesis. The allowable probability of a Type I error in a hypothesis test is called the *level of significance*; the symbol used is α. The critical region is chosen in accord with the level of significance. Each test also has some probability β of Type II error.

	Do not reject H_0	Reject H_0
H_0 is correct	Correct decision	Type I error
H_0 is false (H_1 is true)	Type II error	Correct decision

Figure 1.12 Type I and Type II Errors.

▶ **Example 1.12** We wish to test a hypothesis about the mean of a normal population. In particular, the null hypothesis $H_0 : \mu_X = 20$ is to be tested against the alternative $H_1 : \mu_X \neq 20$. The test has a significance of .05. In other words, the chances of Type I error are limited to 1 in 20. The critical region must be chosen in accord with this requirement; this means that the probability of rejection when the null hypothesis is true must be .05.

Suppose that we use the statistic $\overline{X}$ from a sample of 100 to test the null hypothesis. The sample yields a mean $\overline{X} = 22$ and a sample standard deviation $S_X = 30$. Can we reject the null hypothesis at the .05 level of significance?

Because the sample is greater than 30, we assume that $\sigma_X = S_X = 30$. Under the null hypothesis the mean is 20 and the statistic $\overline{X}$ has a normal distribution, shown in Figure 1.13.

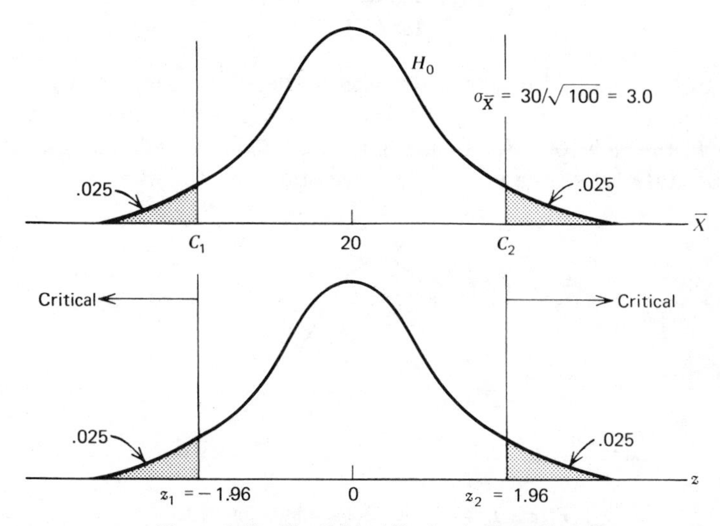

Figure 1.13 The Critical Region in Example 1.12

Values of $\overline{X}$ such that $\overline{X} < C_1$ or $\overline{X} > C_2$ would be unlikely values under the null hypothesis but likely values under the alternative. Therefore the critical region is the shaded area in Figure 1.13. This shaded area must be equal to .05 and we divide the area equally between the two "tails." The critical region can now be found from the corresponding standardized normal distribution:

$$z_1 = \frac{C_1 - 20}{3.0} = -1.96$$

$$\therefore C_1 = 14.12$$

By symmetry $C_2 = 25.88$. Because the value of $\overline{X}$ from the sample ($\overline{X} = 22$) does not fall in the critical region, we have no evidence to reject the null hypothesis at the .05 level of significance. ◀

▶ **Example 1.13** The null hypothesis concerning a normal distribution is $H_0 : \mu_X = 20$, whereas the alternative is $H_1 : \mu_X \geqslant 20$. A sample of 9, used to test the null hypothesis, gives $\overline{X} = 23$ and $S_X = 1.5$. Can we reject the null hypothesis at the .05 level of significance? Given the null hypothesis, the value of t corresponding to $\bar{X} = 23$ is

$$t = \frac{23 - 20}{1.5/\sqrt{9}} = 6.0$$

Because of the nature of H_1, we wish to "lean" towards H_1 if a value of $\bar{X}$ is large. Accordingly, the ciritical region on the corresponding t distribution is to the right of 1.86, as in Figure 1.14. (This is called a "one-tailed" test.) Because 6.0 is greater than 1.86, we reject the null hypothesis. ◀

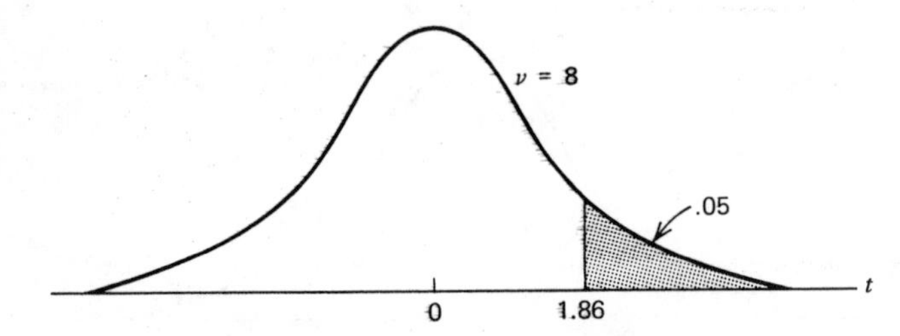

Figure 1.14 Critical Region in Example 1.13

1.13 COVARIANCE, LINEAR DEPENDENCE, AND CORRELATION

Two discrete random variables X_1 and X_2 are *statistically independent* when $\Pr(X_1|X_2)$, the probability that X_1 will occur given that X_2 has occurred, is equal to $\Pr(X_1)$; in other words, when the probability that X_1 will occur is unaffected by the occurrence of any X_2 value. Two continuous random variables X_1 and X_2 are statistically independent when $f(X_1|X_2) = f(X_1)$ or when the probability density of X_1 is unaffected by any X_2 value. It can be shown that for statistical independence it is equivalent to state

$$\Pr(X_1, X_2) = \Pr(X_1) \cdot \Pr(X_2) \tag{1.14}$$

and

$$f(X_1, X_2) = f(X_1) \cdot f(X_2) \tag{1.15}$$

where $\Pr(X_1, X_2)$ is the probability that X_1 and X_2 will occur simultaneously and $f(X_1, X_2)$ is the probability density that X_1 and X_2 will occur simultaneously.

Linear dependence of two variables X_1 and X_2 is measured by the *covariance* given by

$$\mathrm{COV}(X_1, X_2) = E[(X_1 - E(X_1))(X_2 - E(X_2))] \qquad (1.16)$$

Let us first examine covariance intuitively. If X_1 and X_2 tend to move together, in other words, if positive deviations from the means tend to occur together and negative deviations from the means tend to occur together, then $\mathrm{COV}(X_1, X_2)$ will be positive. If negative deviations of X_1 from $E(X_1)$ are accompanied by positive deviations of X_2 from $E(X_2)$ and vice versa, then $\mathrm{COV}(X_1, X_2)$ will be negative. If deviations in X_1 are not related to deviations in X_2, then $\mathrm{COV}(X_1, X_2)$ will tend to be small.

It can be shown[1] that if X_1 and X_2 are statistically independent then $\mathrm{COV}(X_1, X_2) = 0$. The converse is not true. The covariance may be equal to zero but there may be a nonlinear dependence between X_1 and X_2; for example, consider Figure 1.15. There is clearly a relation between X_1 and X_2. In part of the range of X_1 positive deviations of X_1 go with positive deviations of X_2 and in the other part positive deviations go with negative deviations. This could produce $\mathrm{COV}(X_1, X_2) = 0$.

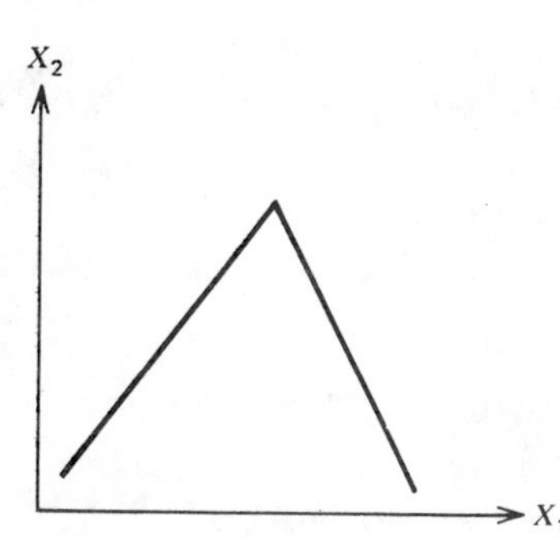

Figure 1.15 A Relationship Between X_1 and X_2

The covariance of two variables X_1 and X_2 can be estimated with a statistic called the *sample covariance*:

$$S_{X_1X_2} = \frac{1}{n-1}\sum_{i=1}^{n}(X_{1i} - \bar{X}_1)(X_{2i} - \bar{X}_2) \qquad (1.17)$$

where the values X_{1i} and X_{2i} occur simultaneously in a sample of n pairs of values. The statistic $S_{X_1X_2}$ is an unbiased estimator of $\mathrm{COV}(X_1, X_2)$.

There is one difficulty with the covariance as a measure of linear dependence. The magnitude of deviations from the mean is dependent on

[1]See Appendix A, Mathematical Note 1.3.

the units in which the random variable is measured. It follows that the magnitude of covariance is dependent on the units used. Because units of measurement are somewhat arbitrary, so is the magnitude of the covariance. This makes it difficult to compare covariances.

The dimensions of $COV(X_1, X_2)$ are (units X_1) · (units X_2). If we divide the covariance by the standard deviation of X_1 in units X_1 and by the standard deviation of X_2 in units X_2, we obtain a dimensionless number called the *correlation coefficient* of X_1 and X_2:

$$\rho_{X_1X_2} = \frac{COV(X_1, X_2)}{SD(X_1) \cdot SD(X_2)} \tag{1.18}$$

It is always true[1] that $|COV(X_1, X_2)| \leqslant |SD(X_1) \cdot SD(X_2)|$; therefore $\rho_{X_1X_2}$ has the range $[-1, 1]$. If X_1 and X_2 have a perfect linear relationship, then $\rho_{X_1X_2}$ will be $+1$ or -1. The negative value occurs if the slope of the line is negative.

In sampling we use a statistic called the *sample correlation coefficient* $r_{X_1X_2}$ to estimate the correlation coefficient of the population.

$$r_{X_1X_2} = \frac{S_{X_1X_2}}{S_{X_1} \cdot S_{X_2}} = \frac{\sum_{i=1}^{n}(X_{1i} - \bar{X}_1)(X_{2i} - \bar{X}_2)}{\sqrt{\sum_{i=1}^{n}(X_{1i} - \bar{X}_1)^2}\sqrt{\sum_{i=1}^{n}(X_{2i} - \bar{X}_2)^2}} \tag{1.19}$$

It is important to remember that $r_{X_1X_2}$ is a statistic and may be used to estimate $\rho_{X_1X_2}$ or to test hypotheses regarding $\rho_{X_1X_2}$.

For small values of n the statistic $r_{X_1X_2}$ has a relatively large variance. In Figure 1.16 is given the density of $r_{X_1X_2}$ for $n = 8$ if $\rho_{X_1X_2}$ is actually zero.

In multiple regression analysis we are interested in all possible relationships among several variables. It is usual to present these relationships, whether they be sample correlations or sample covariances, in matrix

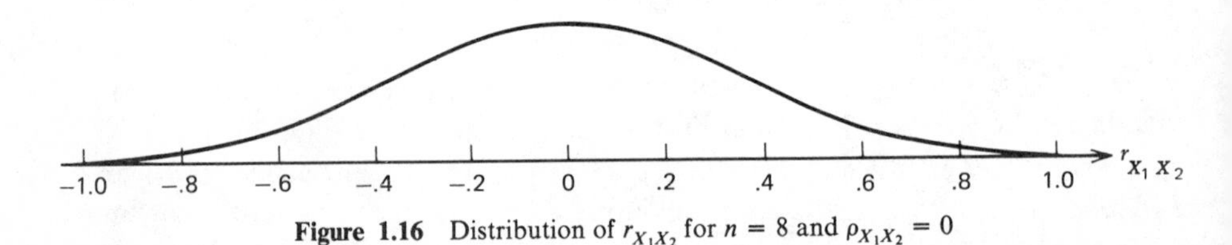

Figure 1.16 Distribution of $r_{X_1X_2}$ for $n = 8$ and $\rho_{X_1X_2} = 0$

[1]This result is given here without proof.

form; for example, for three variables X_1, X_2, and X_3 the covariance matrix would be

$$\begin{bmatrix} S_{X_1X_1} & S_{X_1X_2} & S_{X_1X_3} \\ S_{X_2X_1} & S_{X_2X_2} & S_{X_2X_3} \\ S_{X_3X_1} & S_{X_3X_2} & S_{X_3X_3} \end{bmatrix}$$

It should be noted that in any covariance matrix $S_{X_iX_i} = S_{X_i}^2$, as shown by comparison of (1.17) and (1.8). Because the values in the diagonal (upper left to lower right) are variances, the covariance matrix is sometimes known as the *variance-covariance matrix*. It should also be noted from (1.17) that $S_{X_iX_j} = S_{X_jX_i}$. Some computer programs only print $S_{X_iX_j}$ for $i \geqslant j$, or the diagonal and the lower half of the matrix.

A correlation matrix for three variables is

$$\begin{bmatrix} r_{X_1X_1} & r_{X_1X_2} & r_{X_1X_3} \\ r_{X_2X_1} & r_{X_2X_2} & r_{X_2X_3} \\ r_{X_3X_1} & r_{X_3X_2} & r_{X_3X_3} \end{bmatrix}$$

Because $r_{X_iX_i} = 1$, as verified by (1.19), the diagonal values are always units. Again, $r_{X_iX_j} = r_{X_jX_i}$ so that sometimes only half the matrix is given.

The correlation matrix is easier to interpret, but the covariance matrix gives more information. We can obtain a correlation matrix from the values in a covariance matrix according to (1.19). To recreate a con-variance matrix from a correlation matrix we would, in addition, need the sample standard deviations of the variables.

1.14 LINEAR COMBINATIONS OF VARIABLES

It is sometimes useful to know the distribution of a linear combination of variables. A linear combination, Q, of k independent variables is

$$Q = a_1X_1 + a_2X_2 + \cdots + a_kX_k$$

$$= \sum_{i=1}^{k} a_iX_i \qquad (1.20)$$

where the a_i are constants; for example, Q could be $3X_1 - 4X_2 + X_3$.

The expected value of Q, using operation (3) in Section 1.3,

$$E(Q) = E(a_1X_1 + a_2X_2 + \cdots + a_kX_k)$$
$$= a_1E(X) + a_2E(X_2) + \cdots + a_kE(X_k)$$

$$\therefore \quad E(Q) = \sum_{i=1}^{k} a_iE(X_i) \tag{1.21}$$

Suppose that the variance-covariance matrix of the variables X_i is known. The variance of Q can be found by the formula

$$\text{VAR}(Q) = \sum_{i=1}^{k} \sum_{j=1}^{k} a_i\, a_j \text{COV}(X_i, X_j) \tag{1.22}$$

Although the proof of (1.22) is not given, we demonstrate that operation (3) in Section 1.4 is only a special case of (1.22). By (1.22)

$$\text{VAR}(X_1 + X_2) = \sum_{i=1}^{2} \sum_{j=1}^{2} \text{COV}(X_1, X_2)$$

because $a_1 = a_2 = 1$. Hence

$$\text{VAR}(X_1 + X_2) = \text{COV}(X_1, X_1) + \text{COV}(X_1, X_2)$$
$$+ \text{COV}(X_2, X_1) + \text{COV}(X_2, X_2)$$
$$= \text{VAR}(X_1) + \text{VAR}(X_2) + 2\,\text{COV}(X_1, X_2)$$

Equation 1.22 also applies to sample variances and covariances:

$$S_Q^2 = \sum_{i=1}^{k} \sum_{j=1}^{k} a_i\, a_j S_{X_iX_j} \tag{1.23}$$

▶ **Example 1.14** Suppose $Q = 3X_1 - 4X_2 + X_3$, $S_{X_1}^2 = 4$, $S_{X_2}^2 = 3$, $S_{X_3}^2 = 4$, $S_{X_1X_2} = 1$, $S_{X_1X_3} = 0$, and $S_{X_2X_3} = 1$. Find S_Q^2.

First note that $a_1 = 3$, $a_2 = -4$, and $a_3 = 1$:

$$S_Q^2 = \sum_{i=1}^{3} \sum_{j=1}^{3} a_ia_jS_{X_iX_j}$$
$$= 3(3)S_{X_1X_1} + 3(-4)S_{X_1X_2} + 3(1)S_{X_1X_3}$$

$$+ (-4)3S_{X_2X_1} + (-4)(-4)S_{X_2X_2} + (-4)(1)S_{X_2X_3}$$

$$+ 1(3)S_{X_3X_1} + 1(-4)S_{X_3X_2} + 1(1)S_{X_1X_1}$$

$$= 9S^2_{X_1} + 16S^2_{X_2} + S^2_{X_3} - 24S_{X_1X_2} + 6S_{X_1X_3} - 8S_{X_2X_3}$$

$$= 36 + 48 + 4 - 24 + 0 - 8$$

$$= 56$$

◀

Finally, it should be noted that if $X_1, X_2, \ldots, X_3$ are normally distributed then Q is normally distributed.

chapter 2

Multiple linear regression

2.1 THE MODEL

In many practical statistical problems we wish to be able to explain the values of a variable Y by means of other variables $X_1, X_2, \ldots, X_q$; for example, it may be thought that the demand for product Y is a function of its price X_1, the amount spent on advertising X_2, and the expenditures on product distribution X_3.

The connection between the *dependent variable* Y and the *independent variables* $X_1, X_2, \ldots, X_q$ is often sought in a matrix[1] of observations such as

$$\begin{bmatrix} X_{11} & X_{21} & \cdots & X_{q1} & Y_1 \\ X_{12} & X_{22} & \cdots & X_{q2} & Y_2 \\ \vdots & & & & \\ X_{1n} & X_{2n} & \cdots & X_{qn} & Y_n \end{bmatrix} \tag{2.1}$$

Each row of (2.1) is called a data point. By using the n data points observed we wish to establish a relationship between the dependent variable and the independent variables.

It is often postulated that the relationship is a linear one:

$$Y_R = \beta_0 + \beta_1 X_1 + \beta_2 X_2 + \cdots + \beta_q X_q$$

or

$$Y_R = \beta_0 + \sum_{j=1}^{q} \beta_j X_j \tag{2.2}$$

where $\beta_0, \beta_1, \ldots, \beta_q$ are constants called *regression parameters*. Other names are *regression coefficients* and *partial regression coefficients*. A linear relationship is the popular choice for three reasons: it is easy to work with,

[1]The matrix is used here simply to provide an orderly list. See Appendix B.

it is sometimes a first-order approximation for a nonlinear relationship, and we can often use the unscrupulous argument that there is no reason to believe that the relationship should *not* be linear.

It is useful to obtain a geometrical picture of Y_R. If $q = 1$, Y_R represents a line as in Figure 2.1.

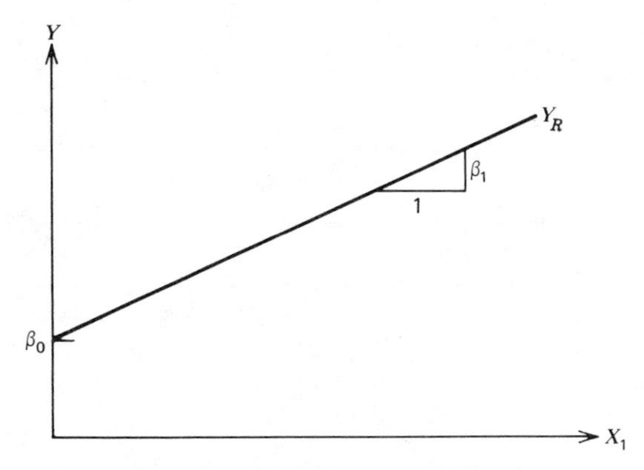

Figure 2.1 Relationship Y_R when $q = 1$.

If $q = 2$, Y_R is a plane in three-dimensional space. In Figure 2.2 ABC is a plane segment of Y_R. The line AB is formed by Y_R cutting the YX_1 plane and the line AC is formed by Y_R cutting the YX_2 plane. With three or more independent variables Y_R is called a hyperplane.

Unfortunately Y_R cannot explain a set of data points such as (2.1) in any except the most unusual circumstances. In the real world data points do not know that they are all supposed to form a line or a plane, hence do not generally do so. The relationship Y_R is not a good model in the sense that it cannot account for the deviation of points from a line or a plane.

This difficulty is overcome by introducing a random variable called the *error term* into the model:

$$Y = Y_R + \epsilon \tag{2.3}$$

and any individual value Y_i is explained by[1]

$$Y_i = \beta_0 + \beta_1 X_{1i} + \beta_2 X_{2i} + \cdots + \beta_q X_{qi} + \epsilon_i \tag{2.4}$$

[1]In the remainder of this book the introduction of a subscript i to a relationship such as (2.2) or (2.3) means that a particular point i is being discussed.

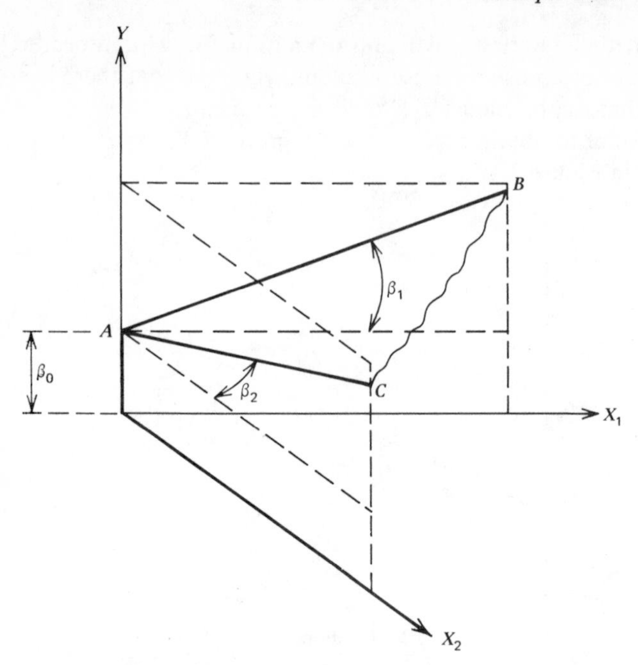

Figure 2.2 Relationship Y_R when $q = 2$.

If any data point does not fall on the line or plane Y_R, there is no longer a problem because we can blame the difference $(Y_i - Y_{Ri})$ on ϵ_i.

In order to facilitate statistical analysis, we make the following assumption about ϵ: ϵ is normally distributed with mean zero and some *constant* standard deviation σ [ϵ is $N(0, \sigma)$]. This assumption is not strictly necessary for many of the estimates and tests that will be devised. Its relaxation is discussed in Chapter 6.

What real-world phenomena can the error term represent? Deviations from a perfect linear relationship can be thought of as the effects of other unspecified variables. It should be noted, however, that we make the assumption that these "error" variables "average out" to zero in the sense that together they have an expected effect of zero.[1]

It is sometimes said that the error term can be used to represent measurement errors in the variables. Only errors in the dependent variable Y can be so represented. Allowing a random error in the X's leads us to a different model.[2]

[1]An error with a constant component, i.e., $E(\epsilon) = C$, simply means that C can be absorbed into β_0. Note that if $E(\epsilon) = 0$ then, by (2.3), $E(Y) = Y_R$.

[2]See Appendix A, Note 2.1.

2.2 ESTIMATING THE PARAMETERS OF THE MODEL

In multiple linear regression analysis we assume that all the values we observe have been generated, in part, by random deviations from the true linear relationship Y_R given in (2.2).

Because of these random deviations, we cannot find Y_R exactly, for we do not know the particular values of ϵ that have occurred. We can, however, estimate the hyperplane Y_R. We label the estimate

$$\hat{Y}_R = b_0 + b_1X_1 + \cdots + b_qX_q \tag{2.5}$$

With due respect for antiquity we use Greek letters for the parameters to be estimated and the corresponding English letters for the estimators. Thus b_0 is an estimator for β_0, b_1 is an estimator for β_1, and so on. Our estimators, called *sample regression coefficients* or *sample partial regression coefficients*,[1] are functions of the data observed and of course, are statistics —random variables, each with its own probability distribution.

The problem is to choose the proper function of sample values to be each estimator. We would prefer linear unbiased estimators with the minimum variance possible. Fortunately the magic of calculus allows us to derive them ([2.1], p. 353). The result (without proof) is as follows: choose that line, plane, or hyperplane that minimizes the sum of the squared deviations of the observed values Y_i from $\hat{Y}_{Ri}$.

Finding the estimators $b_0, b_1, b_2, \ldots, b_q$ is now equivalent to the following mathematical problem:

Find $b_0, b_1, \ldots, b_q$ to minimize

$$\sum_{i=1}^{n} (Y_i - \hat{Y}_{Ri})^2 = \sum_{i=1}^{n} e_i^2 \tag{2.6}$$

The difference $e_i = Y_i - \hat{Y}_{Ri}$ is called a *residual*. It represents the difference of an observed value of the dependent variable from the value "predicted" by the estimated linear relationship. Equation 2.6 can be expanded:

$$\sum_{i=1}^{n} e_i^2 = \sum_{i=1}^{n} [Y_i - (b_0 + b_1X_{1i} + b_2X_{2i} + \cdots + b_qX_{qi})]^2 \tag{2.7}$$

Note that the values X_{ji} are constants as far as the minimizing problem is concerned. They represent actual observations of data.

[1]Because the terms regression coefficient and sample regression coefficient are somewhat cumbersome, the shorter terms "parameter" and "coefficient," respectively, are generally used in this book.

For $q = 1$ the linear relationship could be found by trial and error; for example, there are four data points in Figure 2.3. We could draw a line by eye, measure e_i for $i = 1, 2, 3, 4$ and calculate $e_1^2 + e_2^2 + e_3^2 + e_4^2$. This procedure could be repeated until we were sure that the minimum value of $\Sigma_{i=1}^4 e_i^2$ had been found.

For $q = 2$ the procedure is conceptually just as simple.

Figure 2.4 shows a plane $\hat{Y}_R$ fitted to four points in space. However,

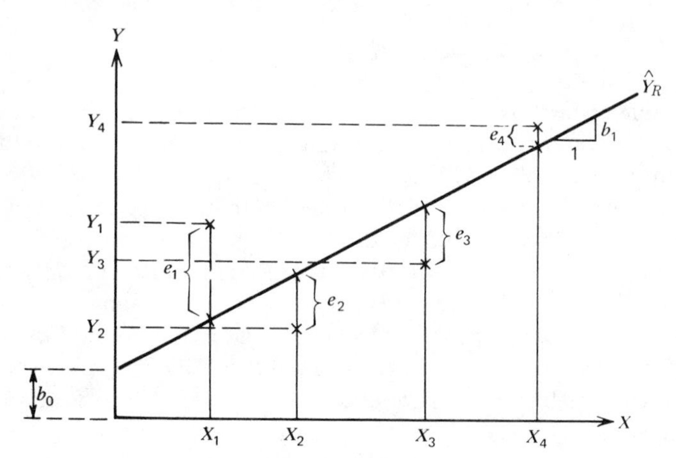

Figure 2.3 Residuals from a line.

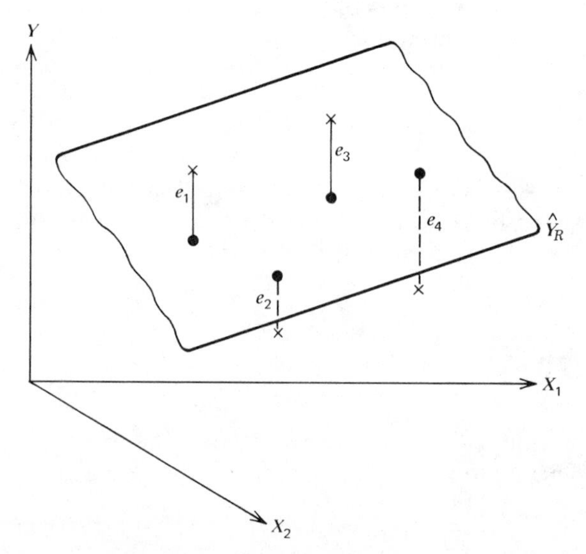

Figure 2.4 Residuals from a plane.

adjusting the position of this plane to minimize Σe_i^2 is more difficult to do graphically.

Fortunately the *least squares* fitting of the relationship $\hat{Y}_R$ may be accomplished with formulas derived from calculus procedures for minimization.[1]

When $q = 1$,

$$b_1 = \frac{\sum_{i=1}^{n} (X_i - \overline{X})(Y_i - \overline{Y})}{\sum_{i=1}^{n} (X_i - \overline{X})^2} \tag{2.8}$$

$$b_0 = \overline{Y} - b_1\overline{X} \tag{2.9}$$

where $\overline{X} = \Sigma_{i=1}^{n} X_i/n$ and $\overline{Y} = \Sigma_{i=1}^{n} Y_i/n$.

The estimators b_0 and b_1 are functions of sample values. Similar statistics can be found[1] for the case with q independent variables: $b_0, b_1, \ldots, b_q$.

It can be shown that each statistic b_i has a normal distribution with expected value β_i (it is unbiased) and a standard deviation σ_{b_i}. Unfortunately this standard deviation is not known but must be estimated from the data.[2] These estimates are $S_{b_0}, S_{b_1}, \ldots, S_{b_q}$; for example, when $q = 1$ (Y_R is a line),

$$S_{b_0}^2 = \frac{\sum_{i=1}^{n} e_i^2}{n-2} \cdot \frac{\sum_{i=1}^{n} X_i^2}{\sum_{i=1}^{n} (X_i - \overline{X})^2}$$

$$S_{b_1}^2 = \frac{\sum_{i=1}^{n} e_i^2}{n-2} \cdot \frac{1}{\sum_{i=1}^{n} (X_i - \overline{X})^2} \tag{2.10}$$

In general the statistic

$$t = \frac{b_i - \beta_i}{S_{b_i}} \tag{2.11}$$

[1]See Appendix A, Note 2.2.

[2]It can be shown that the estimated covariance matrix for the parameter estimators is $(X'X)^{-1} \cdot \Sigma_{i=1}^{n} e_i^2/(n - q - 1)$. The derivation is beyond the scope of this book.

has a *t distribution*. The number of parameters in the linear relationship Y_R is $m = q + 1$. The t distribution in (2.11) has $\nu = n - m$ degrees of freedom.

▶ **Example 2.1** A hypothetical researcher who is also somewhat naïve has a hypothetical problem. He wishes to explain the demand for public transportation to the downtown areas of cities; therefore he gathers data on 20 cities. These data[1] are presented in Table 2.1.

Table 2.1 Observations on 20 Cities

City	X_1	X_2	X_3	X_4	X_5	X_6	Y
1	.73	.94	6.89	1.24	.91	.36	9.33
2	1.05	.99	6.93	.76	1.06	.29	11.56
3	.82	1.02	7.21	1.01	.90	.38	10.32
4	1.13	.85	7.09	1.37	1.02	.28	11.35
5	.89	1.10	7.01	1.14	.95	.31	11.21
6	.86	.92	7.12	1.38	1.04	.34	10.90
7	.97	1.11	6.69	.76	.93	.32	11.82
8	.85	.89	6.91	.97	.91	.29	10.06
9	.83	.85	6.84	1.70	1.02	.29	9.91
10	1.00	.93	6.94	1.11	.95	.29	11.98
11	1.21	1.17	6.95	.99	1.09	.27	13.69
12	1.02	.97	6.92	1.42	1.05	.28	11.44
13	.76	1.25	6.98	1.06	.90	.39	10.58
14	1.30	1.03	7.07	1.33	1.17	.22	14.10
15	1.03	.83	7.08	1.20	.99	.27	10.76
16	1.07	.98	7.04	2.08	1.11	.26	10.59
17	.82	.96	7.02	.86	.72	.32	10.12
18	.73	.91	7.11	1.36	.72	.37	9.36
19	.87	.90	6.98	.57	.75	.32	10.72
20	1.02	1.02	7.10	1.89	1.04	.32	10.92

In Table 2.1

Y = number of passengers per year in the downtown area (in millions)
X_1 = number of stores downtown (in thousands)
X_2 = number of nonretail business offices (in thousands)
X_3 = average family income in the city (in thousands of dollars)
X_4 = fare price index: fare per trip divided by national average
X_5 = population of city (in millions)
X_6 = number of cars per capita

[1]The data were actually produced by $Y = -4 + 5X_1 + 3X_2 + X_3 - X_4 + 2X_5 - X_6 + \epsilon$, where ϵ was randomly generated from a normal distribution with mean 0 and $SD(\epsilon) = \sigma = 0.3$.

The researcher also wishes to estimate the linear relationship (hyperplane):

$$Y_R = \beta_0 + \beta_1 X_1 + \beta_2 X_2 + \beta_3 X_3 + \beta_4 X_4 + \beta_5 X_5 + \beta_6 X_6$$

By running the problem on the BMD03R [2.2] regression program he would obtain the output concerning parameter estimation in Table 2.2.

Table 2.2 BMD03R Printout

INTERCEPT (A VALUE)		1.29071			
VARIABLE	MEAN	STD DEVIATION	REG COEFF	STD ERROR OF REG COEFF	COMPUTED T VALUE
1	.94800	.15813	5.67941	1.94580	2.91881
2	.98100	.11002	3.16294	1.34318	2.35482
3	6.99400	.11731	.25608	1.15942	.22087
4	1.21000	.37811	−.89579	.39040	−2.29454
5	.96150	.12462	1.17651	1.65895	.70919
6	.30850	.04344	−1.87983	6.56232	−.28646
7(Y)	11.03600	1.22282			

The T value column is explained later. Table 2.2 can be translated into our notation as in Table 2.3 (three significant figures are used in Table 2.3)

Table 2.3 Sample Regression Coefficients and Their Standard Errors (from Table 2.2)

i	b_i	S_{b_i}
0	1.29	Not given
1	5.68	1.95
2	3.16	1.34
3	.256	1.16
4	−.896	.390
5	1.18	1.66
6	−1.88	6.56

It is customary to write the estimated linear relationship in the following way:

$$\hat{Y}_R = \underset{}{1.29} + \underset{(1.95)}{5.68X_1} + \underset{(1.34)}{3.16X_2} + \underset{(1.16)}{.256X_3} - \underset{(.390)}{.896X_4} + \underset{(1.66)}{1.18X_5} - \underset{(6.56)}{1.88X_6}$$

The numbers in brackets are the standard errors of the estimators. ◀

▶ **Example 2.2** Find the 90% confidence interval for β_3 in Example 2.1. From Section 1.11 the interval for the $(1 - \alpha) \times 100\%$ confidence level is

$$\hat{I}_{1-\alpha} = \left[b_3 - S_{b_3}(t_{\alpha/2}), b_3 + S_{b_3}(t_{\alpha/2}) \right]$$

Since $1 - \alpha = .9$, $\alpha = .1$. With $n - m$ or $20 - 7$ degrees of freedom $t_{\alpha/2}$ or $t_{.05} = 1.771$ (from Table 2 in Appendix C). Then

$$\hat{I}_{.9} = [.256 - 1.16(1.77), .256 + 1.16(1.77)]$$

$$= [-1.80, 2.31]$$

or $[-1.80 \leqslant \beta_3 \leqslant 2.31]$ at the 90% confidence level. ◀

▶ **Example 2.3** Test the hypothesis $H_0 : \beta_4 = 0$ against $H_1 : \beta_4 \neq 0$ at the .01 level of significance. The relevant distributions for the case in which the null hypothesis is true are given in Figure 2.5. The t value of $b_4 = -.896$ under the null hypothesis is $t = (-.896 - 0)/.39 = -2.30$, as printed out by the BMD03R program in Table 2.2. It does not fall in the critical region, hence we cannot reject the null hypothesis that $\beta_4 = 0$, at least not at the .01 level of significance. ◀

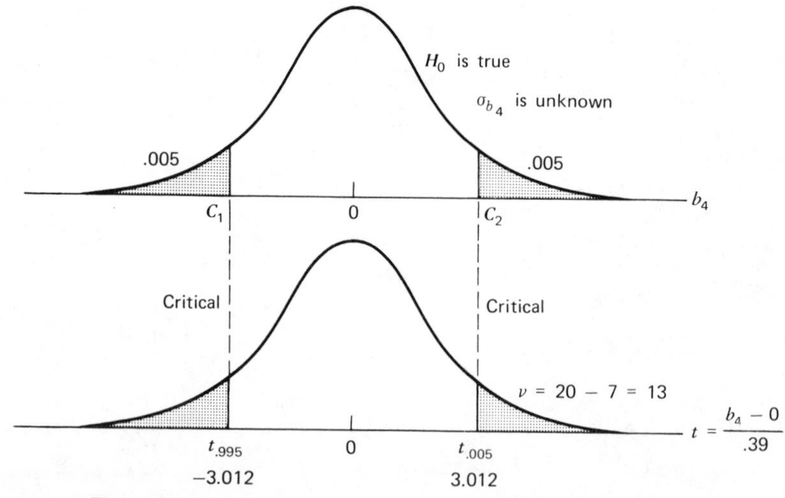

Figure 2.5 Distribution of b_4 under the null hypothesis $H_0 : \beta_4 = 0$.

Note that the confidence intervals and tests for regression parameters were calculated one at a time; for example, if in some regression we have 95% confidence in an interval for β_1, and 95% confidence in an interval for β_2, it does not mean that we have 95% confidence that both intervals are simultaneously correct. Simultaneous intervals and tests are discussed in connection with analysis of variance models in Section 7.7. See also Note 2.3 in Appendix A.

2.3 PARTITIONING SUMS OF SQUARES

The least squares concept not only provides a method for estimating the parameters of an assumed linear relationship but is also the basis for many other tests and interpretations.

We can consider sums of squared deviations as measures of how much a variable differs from a specified value; for example, the sum of squared deviations from the mean value of observed Y values is a measure of how much the dependent variable varies from its own mean. We now wish to see how much of this variation can be "accounted for" by the fitted linear relationship. If a great deal of the variation in Y is explained by the model, we have confidence in the model. The following identity[1] is fundamental to much of the analysis to follow:

$$\sum_{i=1}^{n} (Y_i - \bar{Y})^2 = \sum_{i=1}^{n} (\hat{Y}_{Ri} - \bar{Y})^2 + \sum_{i=1}^{n} (Y_i - \hat{Y}_{Ri})^2 \qquad (2.12)$$

Consider first the term $\Sigma(Y_i - \bar{Y})^2$. The distance $(Y_i - \bar{Y})$ is the deviation of an observed value Y_i from the mean of all observed values. The term $\Sigma(Y_i - \bar{Y})^2$ is therefore a measure of the variability of the dependent variable. According to the identity (2.12), this variation is ascribed to two causes: to deviations of the linear relationship from the mean and to the presence of residuals.

For every set of observations $(X_{1i}, X_{2i}, \ldots, X_{qi})$ there is a point $\hat{Y}_{Ri}$. The distance $(\hat{Y}_{Ri} - \bar{Y})$ can be attributed to the linear relationship's "attempt" to explain deviations of Y from $\bar{Y}$. In Figure 2.6 the heavy lines represent such distances for the case in which $q = 1$ (straight line relationship).[2] The term $\Sigma(\hat{Y}_{Ri} - \bar{Y})^2$ gives the model's contribution to explaining the variation of Y about its mean.

[1]See Appendix A, Note 2.4 for a proof.

[2]When $\hat{Y}_R$ is a hyperplane, $(\hat{Y}_{Ri} - \bar{Y})$ is that deviation of the Y value from the hyperplane $\bar{Y}$ that is predicted by $\hat{Y}_R$.

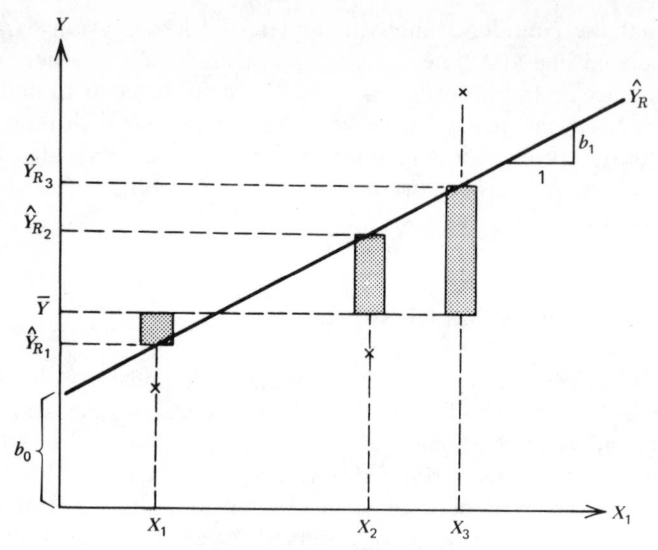

Figure 2.6 Variation "explained" by Y_R.

The model cannot fully explain all variations in Y, and the plane $\hat{Y}_R$ usually cannot pass through all the data points. The deviation $(Y_i - \hat{Y}_{Ri})$ is the "vertical" distance of the point i from the plane $\hat{Y}_R$. Figure 2.7 again

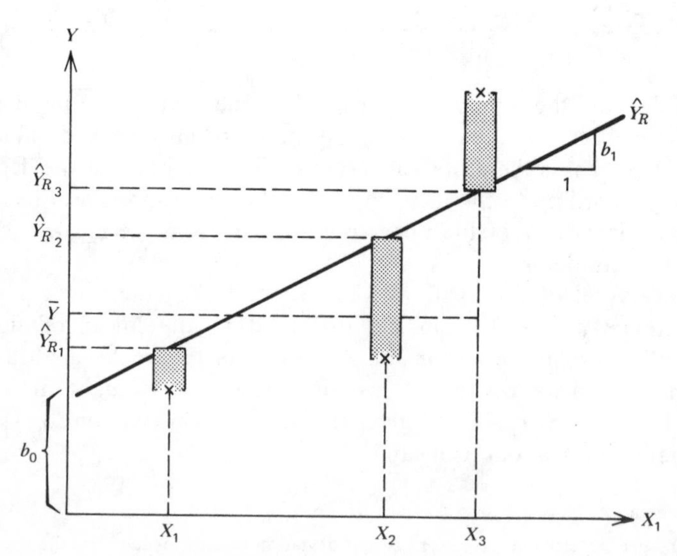

Figure 2.7 Variation "unexplained" by $\hat{Y}_R$.

illustrates the case[1] in which $q = 1$. The heavy lines in Figure 2.7 are the distances $(Y_i - \hat{Y}_{Ri})$. In general, the term $\Sigma(Y_i - \hat{Y}_{Ri})^2$ shows the sum of squared residuals or the variation unexplained by $\hat{Y}_R$.

In summary, the partitioning of (2.12) yields

$$\text{SS (sum of squares) about the mean} = \text{SS due to regression} + \text{SS about regression} \tag{2.13}$$

or, alternatively,

$$\text{total SS} = \text{explained SS} + \text{unexplained SS} \tag{2.14}$$

2.4 THE ANALYSIS OF VARIANCE TABLE AND MEAN SQUARES

It is useful to consider the degrees of freedom possessed by each of the sums of squares in (2.12). *Degrees of freedom* refer to the number of independent pieces of information involving the dependent variable Y needed to compile the sum of squares.

Consider first the sum of squares about the mean. Only $(n - 1)$ of the numbers $(Y_i - \bar{Y})$ are needed because, knowing them, we can find the nth. This is true because the sum of all $(Y_i - \bar{Y})$ must equal zero:

$$\begin{aligned}\sum_{i=1}^{n} (Y_i - \bar{Y}) &= \sum_{i=1}^{n} Y_i - n\bar{Y} \\ &= n\bar{Y} - n\bar{Y} \\ &= 0\end{aligned}$$

It follows that $\Sigma(Y_i - \bar{Y})^2$ has $n - 1$ degrees of freedom. This result could have been obtained by the rationalization that one function of the Y_i's, namely $\bar{Y}$, was used in calculating the sum of squares; hence one degree of freedom was lost.

This rationalization is handy in finding the degrees of freedom of the sum of squares about regression. In $\Sigma(Y_i - \hat{Y}_{Ri})^2$, the function $\hat{Y}_{Ri}$ is completely determined by the m(or $q + 1$) values $b_0, b_1, \ldots, b_q$. This sum has therefore lost m degrees of freedom and has $n - m$ degrees remaining.

To find the degrees of freedom of the sum of squares due to regression we shall be somewhat sneaky and note that the degrees of freedom on the left-hand side of (2.12) must be equal to the degrees of freedom on the

[1]In the general case the distances $(Y_i - \hat{Y}_{Ri})$ are the distances of the data points from the regression hyperplane $\hat{Y}_R$ (measured parallel to the Y axis).

right-hand side:

$$n - 1 = \text{degrees of freedom for } \Sigma\left(\hat{Y}_{Ri} - \overline{Y}\right)^2 + n - m$$

By solving the equation we find that the sum of squares due to regression has $m - 1$ degrees of freedom. The sums of squares and their degrees of freedom are usually summarized in what is called an analysis of variance table (Table 2.4).

Table 2.4 Analysis of Variance

Source of variation	Sum of squares	Degrees of freedom	Mean square
Due to regression	$\Sigma(\hat{Y}_{Ri} - \overline{Y})^2$	$m - 1$	$\dfrac{\Sigma(\hat{Y}_{Ri} - \overline{Y})^2}{(m - 1)}$
About regression	$\Sigma(Y_i - \hat{Y}_{Ri})^2$	$n - m$	$\dfrac{\Sigma(Y_i - \hat{Y}_{Ri})^2}{n - m}$
Total sum of squares	$\Sigma(Y_i - \overline{Y})^2$	$n - 1$	$\dfrac{\Sigma(Y_i - \overline{Y})^2}{(n - 1)}$

A *mean square* is obtained by dividing a sum of squares by the appropriate degrees of freedom. The three mean squares in Table 2.4 are valuable statistics. Their importance lies in the fact that each can be used to estimate σ^2, the variance of the error term.

Consider first the statistic $[\Sigma(Y_i - \overline{Y})^2]/(n - 1)$, or the *total mean square*. Suppose that in the true relationship Y_R, $\beta_1 = \beta_2 = \cdots = \beta_q = 0$. As we shall see, this peculiar case is important for theoretical reasons. In this case $Y_i = \beta_0 + \epsilon_i$ and $\text{VAR}(Y) = \text{VAR}(\epsilon) = \sigma^2$. It follows from (1.8) that $[\Sigma(Y_i - \overline{Y})^2]/(n - 1)$ is an unbiased estimator for σ^2 if this condition holds.

The *mean square about regression* $[\Sigma(Y_i - \hat{Y}_{Ri})^2]/(n - m)$ is a statistic that is also an unbiased estimator for σ^2, this time unconditionally.

Recall that

$$e_i = (Y_i - \hat{Y}_{Ri})$$

and

$$\epsilon_i = (Y_i - Y_{Ri})$$

It seems reasonable to estimate the variance of the ϵ_i's with the sample variance of the e_i's.

We first note that $\bar{e}$ is always zero.[1] The average squared deviation of the e_i's from the mean is

$$\frac{\Sigma(e_i - \bar{e})^2}{n} = \frac{\Sigma e_i^2}{n}$$

$$= \frac{\Sigma(Y_i - \hat{Y}_{Ri})^2}{n}$$

This statistic, however, would be a biased estimator of σ^2. Use of the divisor $n - m$ instead of n (where m is the number of parameters in Y_R) can be shown to give an unbiased estimator.

Incidentally, the mean square about regression is sometimes called the *mean square due to error* (MSE) and its square root, the *standard error of the estimate*.

Finally, the *mean square due to regression* $[\Sigma(\hat{Y}_{Ri} - \bar{Y})^2]/(m - 1)$ is also an unbiased estimator of σ^2; the condition again is that $\beta_1 = \beta_2 = \cdots = \beta_q = 0$. The proof is beyond the scope of this book. This mean square is a measure of the deviation of the fitted plane (line) $\hat{Y}_R$ from the "horizontal" plane (line) $\bar{Y}$.

It is intuitively reasonable that if the condition $\beta_1 = \beta_2 = \cdots = \beta_q = 0$ is actually not true the mean square due to regression will overestimate σ^2. This property, along with the properties of the previously discussed mean squares, will be useful in the next two sections. First, however, let us consider an example of a computer printout for an analysis of variance table.

▶ **Example 2.4** The data in Example 2.1 were run on the BMD03R program. The resulting analysis of variance is presented in Table 2.5. There were 20 data points, and a linear relationship $\hat{Y}_R$ with six independent variables (seven parameters) was being fitted. Thus

$$n = 20$$

$$q = 6$$

$$m = q + 1 = 7$$

The degrees of freedom in Table 2.5 can now be checked from Table 2.4. The total mean square was not printed; it is 28.41068/19.

[1]See Appendix A, Note 2.5.

Table 2.5 Analysis of Variance for Example 2.4

VARIANCE OF ESTIMATE .21035
STD ERROR OF ESTIMATE .45864
ANALYSIS OF VARIANCE FOR THE MULTIPLE LINEAR REGRESSION

SOURCE OF VARIATION	DF	SUM OF SQUARES	MEAN SQUARES
DUE TO REGRESSION	6	25.67611	4.27935
DEVIATION ABOUT REGRESSION	13	2.73457	.21035
TOTAL	19	28.41068	

If we had to judge how well the linear relationship we found explains the variation in Y, we might look at the relative sizes of the sum of squares due to regression and the sum about regression. Because the squares about the regression produce a much smaller number in this example, we might venture to say that the regression produced a "good" fit. Such a comparison is made in the next section (with due attention paid to degrees of freedom). ◀

2.5 THE F TEST

It is theoretically true that any estimated linear relationship $\hat{Y}_R$ may be due to chance; for example, we may have 20 observations on six variables. Suppose we then generate 20 "dependent" values entirely by random numbers and with no relation to our six independent variables. A regression run on this data almost certainly will not produce $b_1 = b_2 = b_3 = \cdots = b_q = 0$. In other words, chance can generate a hyperplane $\hat{Y}_R$ that is not identical with $\overline{Y}$. Then how can we assure ourselves that a relationship $\hat{Y}_R$ fitted to real data is not due to chance? The answer is that we can never be sure but we can get an idea of how likely or unlikely such a chance generation of $\hat{Y}_R$ would be.

The standard hypothesis testing method is used. We assume the worst with the null hypothesis H_0:

$$H_0 : \beta_1 = \beta_2 = \cdots = \beta_q = 0$$

The alternative hypothesis H_1 is

$$H_1 : \text{not all } \beta_i \, (i > 0) \text{ are equal to zero.}$$

To be able to reject H_0 when warranted we need a statistic with a known distribution when H_0 is true. Then "unlikely" values of the statistic calculated from the data will lead to the rejection of H_0.

It can be shown that if H_0 is true the statistic

$$F = \frac{\text{mean square due to regression}}{\text{mean square about regression}}$$

or

$$F = \frac{\left[\Sigma\left(\hat{Y}_{Ri} - \overline{Y}\right)^2\right] / (m - 1)}{\left[\Sigma(Y_i - \hat{Y}_{Ri})^2\right] / (n - m)} \tag{2.15}$$

has an F distribution with $\nu_1 = m - 1$ degrees of freedom for the numerator and $\nu_2 = n - m$ degrees of freedom for the denominator. If the null hypothesis is true, and for values of F greater than one, larger values of F become increasingly more unlikely. Why should this be so?

One explanation is provided by looking at the mean squares in (2.15) as estimators of $\sigma^2 = \text{VAR}(\epsilon)$. As seen in the preceding section, the denominator is an unbiased estimator for σ^2 regardless of the values of β_1, $\beta_2, \ldots, \beta_q$. The numerator is an unbiased estimator for σ^2 only if the null hypothesis is true; if it is not, then that statistic will tend to be large. A large value of F casts doubts on the null hypothesis.

Another way of rationalizing the behavior of F is to use the alternative terminology in (2.14).

$$F = \frac{\text{explained SS}/(m - 1)}{\text{unexplained SS}/(n - m)}$$

$$= \frac{\text{explained MS}}{\text{unexplained MS}}$$

If the null hypothesis were true, it would be unlikely that the explained MS would be much greater than the unexplained MS; for instance, if the estimators $b_1, b_2, \ldots, b_q$ were all exactly zero, the explained SS would be exactly zero ($\hat{Y}_R = \overline{Y}$). Small values of $b_1, b_2, \ldots, b_q$ should produce a small explained SS.

Table 3 in Appendix C gives critical regions of F at the 1 and 5% levels of significance. If we choose the 1% level of significance, for example, we are saying that we will tolerate a probability of .01 of rejecting the null hypothesis ($H_0 : \beta_1 = \beta_2 = \cdots = \beta_q = 0$) when it is true. If a calculated value of F is greater than the critical value found in Table 3, we reject the null hypothesis and thereby reject the notion that our regression slopes are different from zero purely by chance.

▶ **Example 2.5** Table 2.6 reproduces Table 2.5 which was calculated for the data of Example 2.1, but this time the computed F value is included.

Table 2.6 Analysis of Variance as in Table 2.5 but with the F Value Included

VARIANCE OF ESTIMATE .21035
STD ERROR OF ESTIMATE .45864
ANALYSIS OF VARIANCE FOR THE MULTIPLE LINEAR REGRESSION

SOURCE OF VARIATION	DF	SUM OF SQUARES	MEAN SQUARES	F VALUE
DUE TO REGRESSION	6	25.67611	4.27935	20.3438
DEVIATION ABOUT REGRESSION	13	2.73457	.21035	
TOTAL	19	28.41068		

We can check the F value by calculating the appropriate ratio of mean squares:

$$F = \frac{\text{mean square due to regression}}{\text{mean square about regression}}$$

$$= \frac{4.27935}{.21035}$$

$$= 20.344$$

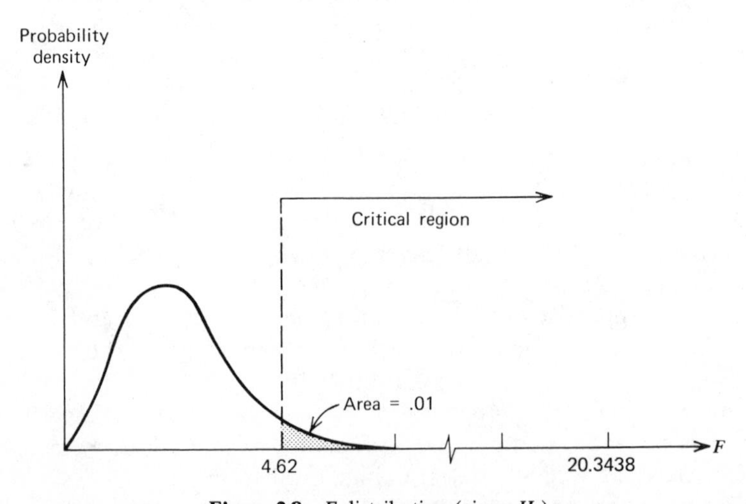

Figure 2.8 F distribution (given H_0).

Suppose that we choose to test the hypothesis

$$H_0 : \beta_1 = \beta_2 = \beta_3 = \beta_4 = \beta_5 = \beta_6 = 0$$

at the .01 level of significance. $F_{.01} = 4.62$ in Table 3 when $\nu_1 = 7 - 1 = 6$ and $\nu_2 = 20 - 7 = 13$. As shown in Figure 2.8, the F value of the fitted regression falls in the critical region and the null hypothesis is rejected. ◀

2.6 THE COEFFICIENTS OF DETERMINATION AND CORRELATION

Our model has allowed us to consider some of the variation of Y as due to a linear relationship Y_R and some as due to an error term ϵ. A measure of the relative importance of each "source" of variation is the coefficient of determination. The *coefficient of determination* is the ratio of the sum of squares due to regression (explained by regression) to the sum of squares about $\overline{Y}$. The common symbol used for this coefficient is R^2.

$$R^2 = \frac{\text{SS due to (explained by) regression}}{\text{SS about the mean}}$$

From (2.13)

$$R^2 = \frac{\text{SS about the mean} - \text{SS about regression}}{\text{SS about the mean}}$$

$$= 1 - \frac{\text{SS about regression}}{\text{SS about mean}}$$

$$R^2 = 1 - \frac{\Sigma(Y_i - \hat{Y}_{Ri})^2}{\Sigma(Y_i - \overline{Y})^2} \tag{2.16}$$

It is informative to examine two extreme cases. If the fit of the linear relationship $\hat{Y}_R$ is perfect, that is, if all the points fall on $\hat{Y}_R$, then $\Sigma(Y_i - \hat{Y}_{Ri})^2 = 0$ and $R^2 = 1$. If, on the other hand, the variation from the linear relationship is nearly as large as the variation about the mean of Y, R^2 approaches zero. In the latter case, of course, the regression has little explanatory value. The value of R^2 is therefore a measure of the "explanatory" power of the regression. It can be shown that when $q = 1$ (there is only one independent variable) $R^2 = r^2_{YX_1}$ where $r^2_{YX_1}$ is defined according to (1.19).

It should be remembered that R^2 is a statistic, hence that the particular value it takes is governed by chance. When samples are small in relation to the number of parameters being fitted, it is possible to get a large value of R^2 even when no linear relationship exists ($\beta_1 = \beta_2 = \cdots = \beta_q = 0$).

If we wish to estimate the fraction of the variance of Y that is attributable to the linear relationship Y_R, it is customary to use $\bar{R}^2$, where

$$\bar{R}^2 = 1 - \frac{\text{mean square about regression}}{\text{mean square about the mean}}$$

$$\bar{R}^2 = 1 - \frac{\left[\Sigma(Y_i - \hat{Y}_{Ri})^2\right]/(n-m)}{\left[\Sigma(Y_i - \bar{Y})^2\right]/(n-1)} \tag{2.17}$$

The coefficient $\bar{R}^2$ is called the *adjusted coefficient of determination* or the *corrected coefficient of determination* because it is adjusted for degrees of freedom. The mean square about regression is an unbiased estimator for σ^2, the variance of the error term, and the mean square about the mean is an unbiased estimator for VAR(Y) (at the given values of $X_1, X_2, \ldots, X_q$). However, $\bar{R}^2$ may sometimes have a negative value, which may be disturbing to some.

The square root of R^2 is called the *coefficient of multiple correlation*. It is difficult to attach a useful meaning to the sign of this coefficient, hence R is usually given as a positive number between zero and one. It is also a measure of the explanatory power of the regression but lacks the intuitive appeal of R^2. Some researchers, noting that the square root of a number between zero and one is larger than the number itself and wishing to present their work in the best possible light, cleverly prefer to talk about R rather than R^2.

The statistic R^2 is closely related to F and can in fact be calculated from the F value. The reverse is also true.

From (2.15)

$$F = \frac{n-m}{m-1} \cdot \frac{\Sigma(\hat{Y}_{Ri} - \bar{Y})^2/\Sigma(Y_i - \bar{Y})^2}{\Sigma(Y_i - \hat{Y}_{Ri})^2/\Sigma(Y_i - \bar{Y})^2}$$

By (2.16) and because $R^2 = \Sigma(\hat{Y}_{Ri} - \bar{Y})^2/\Sigma(Y_i - \bar{Y})^2$

$$F = \frac{n-m}{m-1} \cdot \frac{R^2}{1 - R^2} \tag{2.18}$$

It can also be shown that

$$R^2 = \frac{F(m-1)/(n-m)}{1 + F(m-1)/(n-m)} \tag{2.19}$$

When R^2 is zero, F is also zero [from (2.18)]. As R^2 approaches one, F aproaches infinity. It follows that we can use R^2 to test the null hypothesis $H_0 : \beta_1 = \beta_2 = \cdots = \beta_q = 0$. A critical value of R^2 exists at, say, the .05 level of significance. Tables of R^2 are not readily available, but we can use an F table, find $F_{.05}$, and then use (2.19) to find $R^2_{.05}$.

▶ **Example 2.6** For the data in Example 2.1 the BMD03R program prints out:

COEFFICIENT OF DETERMINATION	.9037
MULTIPLE CORR COEFFICIENT	.9507

We can check that $\sqrt{R^2} = \sqrt{.9037} = .9507$.

Also, from Table 2.6,

$$R^2 = \frac{\text{SS due to regression}}{\text{total SS}}$$

$$= \frac{25.67611}{28.41068}$$

$$= .9037$$

or

$$R^2 = 1 - \frac{\text{SS about regression}}{\text{total SS}}$$

$$= 1 - \frac{2.73457}{28.41068}$$

$$= 1 - .09625$$

$$= .9037$$

or, using (2.19) and $F = 20.3438$,

$$R^2 = \frac{20.3438(7-1)/(20-7)}{1 + 20.3438(7-1)/(20-7)}$$

$$= \frac{9.3894}{1 + 9.3894}$$

$$= .9037$$

As shown in Example 2.5, $F_{.01} = 4.62$;

$$\therefore R^2_{.01} = \frac{4.62(6/13)}{1 + 4.62(6/13)}$$

$$= .681$$

Because $R^2 = .9037$, we can reject $H_0 : \beta_1 = \beta_2 = \cdots = \beta_6 = 0$ at the .01 level of significance.

As a final exercise we find $\bar{R}^2$ by using (2.17):

$$\bar{R}^2 = 1 - \frac{2.73457/13}{28.41068/19}$$

$$= 1 - \frac{.21035}{1.4953}$$

$$= .8593$$

◀

EXERCISES

2.1 A geographer developed a computer program to locate facilities (such as health care centers). In general, the problem considered was how best to choose the locations of X_1 facilities from X_2 possible sites ($X_2 > X_1$). In each problem there were X_3 demand centers; a demand center could be a community such as a town or an artificial entity such as an electoral district.

The program considered transportation and building costs and the convenience of the populations in the demand centers in choosing the best set of locations for the X_1 facilities. It was an iterative process; in other words, a set of locations was chosen and then improved, step by step, until the best solution was reached. The geographer wished to prove that his program was capable of handling problems of a practical size. He expected that computer time would increase with the number of potential sites (X_2), the number of facilities to be located (X_1), and the number of demand centers (X_3). He hoped that his program would not take too much computer time, hence be too expensive to solve practical-sized problems.

Because he held a large research grant, he sent his assistant to 30 population centers in which problems of this kind were being considered or had recently been considered. Having obtained 30 observations of X_1, X_2, and X_3, as well as rough estimates of costs and covenience measures, he solved the 30 problems with his program.

The geographer believed that X_1, X_2, and X_3 were the variables that mainly determined Y, the computer execution time in seconds. He was satisfied to leave all other determinants of computer time to the error term. Table 2.7 presents his data.

(a) Use a prepared computer multiple regression program to obtain a fitted linear relationship with Y as the dependent variable and X_1, X_2 and X_3 as the independent variables.

(b) If the program used in (a) printed out variable sample standard deviations and a correlation matrix, calculate the covariance of variables X_2 and X_3. If a covariance matrix was printed out, use it to calculate the sample correlation coefficient of variables X_2 and X_3.

(c) Write out the fitted linear relationship and place the standard errors of the coefficients below the coefficients in brackets. What do the standard errors measure?

Table 2.7 Data for Exercise 2.1

Observation number	Number of locations (X_1)	Number of potential sites (X_2)	Number of demand centers (X_3)	Central processor execution time (seconds)
1	5	10	96	105
2	10	22	107	290
3	6	11	97	98
4	4	9	83	85
5	7	15	107	168
6	4	12	66	106
7	5	8	105	98
8	8	18	103	191
9	7	15	107	151
10	6	15	82	171
11	3	9	58	95
12	8	23	65	337
13	8	18	108	236
14	3	8	68	110
15	2	11	35	110
16	6	13	82	135
17	5	11	90	134
18	7	17	78	197
19	6	16	72	186
20	4	15	47	149
21	1	4	53	63
22	5	12	84	101
23	4	9	84	101
24	3	6	75	72
25	6	17	66	173
26	9	16	141	174
27	4	8	90	78
28	6	11	96	127
29	4	12	69	105
30	6	11	98	175

(d) Test the hypothesis $H_0 : \beta_1 = 0$ against the hypothesis $H_1 : \beta_1 > 0$ at the .05 level of significance by using a t test.

(e) Test each hypothesis $H_0 : \beta_j = 0$ separately against the alternative $H_1 : \beta_j \neq 0$ at the .01 level of significance by using t tests.

(f) Find the 90% confidence intervals for β_2 and β_3.

2.2 For the computer run of Exercise 2.1:

(a) Check the value for the coefficient of determination (R^2) by using the appropriate ratio of sums of squares. Check the given value of R, the multiple correlation coefficient. Explain the meaning of R^2 and R.

(b) Calculate or check the value of $\bar{R}^2$ by using the appropriate mean squares.

(c) Verify the F value by

(1) computing it as a ratio of sums of squares adjusted by the appropriate degrees of freedom;

(2) computing it from the coefficient of determination.

(d) From the appropriate tables find the critical values for F at the 5 and 1% levels of significance. Can you reject the hypothesis that all the regression coefficients except β_0 are jointly 0?

(e) Is the sign on the multiple correlation coefficient always positive? Why?

2.3 An economics student made a classroom presentation of a regression by using expenditure on food as a dependent variable and total consumption expenditure as an independent variable. He was asked what effect errors in recording or estimating each type of expenditure would have on the validity of his model. Discuss.

2.4 Using the computer run in Exercise 2.1, find the 90% joint or simultaneous confidence intervals for β_2 and β_3 (the material in Appendix A, Note 2.3 is required to answer the question). Compare the answer with that in Exercise 2.1 (f). Why is there a difference?

REFERENCES

1. Jan Kmenta, *Elements of Econometrics*, New York: Macmillan, 1971.
2. *Biomedical Computer Programs*, Health Sciences Computing Facility, Department of Biomathematics, School of Medicine, University of Los Angeles, University of California Press, January 1973.

chapter 3

Interpreting multiple linear regression

3.1 MULTICOLLINEARITY AND OMITTED VARIABLES

One of the chief causes of misinterpretation and misuse of regression is a villain called multicollinearity. *Multicollinearity* is said to exist when any independent (sometimes called *exogenous*) variable X_j is correlated with another independent variable or with a linear combination of other independent variables. Multicollinearity is common and even inevitable in much of the data in fields like sociology, economics, business, and geography. This is true because in these fields it is not generally possible to choose values for the independent variables; one must use the data available. In economics, as an example, data are virtually notorious for multicollinearity.

Correlation in the independent variables causes three main problems:

1. The standard errors of the regression coefficients are increased. In the extreme case in which one independent variable can be expressed as a linear function of others, that is, $X_j = d + \Sigma_{i \neq j} g_i X_i$, in which d and the g_i's are constants, and at least one of the g_i's is nonzero, no regression coefficients can be found.[1]
2. As the extreme case above is approached, computational difficulties arise.
3. The omission of variables may result in biased estimators for the regression parameters of the remaining variables if the missing variables are correlated with those remaining.

Turning to the first problem, consider the two-variable case:

$$\hat{Y}_R = b_0 + b_1 X_1 + b_2 X_2 \tag{3.1}$$

[1]This case is sometimes called *perfect* or *extreme multicollinearity*.

It can be shown that

$$S_{b_1} = \left(\frac{\Sigma e_i^2}{n-3} \right)^{1/2} \cdot \left(\Sigma (X_{1i} - \bar{X}_1)^2 (1 - r^2_{X_1 X_2}) \right)^{-1/2} \quad (3.2)$$

and

$$S_{b_2} = \left(\frac{\Sigma e_i^2}{n-3} \right)^{1/2} \cdot \left(\Sigma (X_{2i} - \bar{X}_2)^2 (1 - r^2_{X_1 X_2}) \right)^{-1/2} \quad (3.3)$$

where

$$r_{X_1 X_2} = \frac{\Sigma \left[(X_{1i} - \bar{X}_1)(X_{2i} - \bar{X}_2) \right]}{\left[\Sigma (X_{1i} - \bar{X}_1)^2 \cdot \Sigma (X_{2i} - \bar{X}_2)^2 \right]^{1/2}}$$

The coefficient $r_{X_1 X_2}$ is, of course, a measure of the correlation of variables X_1 and X_2 in our sample. Note that if $r_{X_1 X_2} = 1$, then both standard errors S_{b_1} and S_{b_2} will be infinitely large. It can be shown similarly, that in general, when there are more than two variables, multicollinearity increases the standard errors of the coefficients.[1]

As the extreme case of a perfect linear relationship among independent variables is approached, computer programs may produce serious errors due to rounding. Computer programs, however, have various ways of warning the user against this danger (e.g., see the footnote on tolerance in Table 4.4).

It should be stressed, however, that when multicollinearity is present all estimators for the parameters of Y_R (the sample regression coefficients) are unbiased; the expected value of a coefficient is equal to the corresponding parameter values; for example, $E(b_2) = \beta_2$.

Yet, paradoxically, the pairwise correlations aspect of multicollinearity is often a cause of biased parameter estimators. If a variable X_j correlated with another X_i is omitted from $\hat{Y}_R$, the estimator b_i may be biased.

As an illustration, consider again (3.1), the case in which $q = 2$.

It can be shown ([2], pp. 288–289) that if the data for X_2 is dropped and

[1]When there are more than two independent variables it is possible to have even extreme multicollinearity [e.g., some cases when, say, $X_2 = X_3 + X_4$ ([1], page 384)] without having particularly high correlations among any pairs of independent variables. The examination of the correlations matrix for the independent variables does not therefore necessarily reveal the degree of multicollinearity.

the relationship $\hat{Y}_R = b_0' + b_1'X_1$ is estimated

$$E(b_1') = \beta_1 + \beta_2 \frac{\Sigma(X_{1i} - \bar{X}_1)(X_{2i} - \bar{X}_2)}{\Sigma(X_{1i} - \bar{X}_1)^2} \tag{3.4}$$

or, if $\hat{Y}_R = b_0'' + b_2''X_2$ is estimated by dropping X_1 data,

$$E(b_2'') = \beta_2 + \beta_1 \frac{\Sigma(X_{1i} - \bar{X}_1)(X_{2i} - \bar{X}_2)}{\Sigma(X_{2i} - \bar{X}_2)^2} \tag{3.5}$$

The first point to remember is that in (3.4) and (3.5) the X_{ji}'s are constants; we are considering the expected values of b_1 and b_2 for a given set of independent variables. The estimators b_1' and b_2'' will be unbiased when $E(b_1') = \beta_1$ and $E(b_2'') = \beta_2$. As can be seen in (3.4) the estimator b_1' will be unbiased if one or both of two conditions hold:

$$\beta_2 = 0$$

and/or

$$S_{X_1X_2} = \frac{\Sigma(X_{1i} - \bar{X}_1)(X_{2i} - \bar{X}_2)}{n - 1} = 0$$

Similarly, b_2'' will be an unbiased estimator for β_2 if $\beta_1 = 0$ and/or $S_{X_1X_2} = 0$.

In general, if we drop a variable j, the remaining regression coefficients will be unbiased estimators if the variable j had no effect on Y_R(i.e., $\beta_j = 0$) and/or the correlations of variable j with all the remaining variables are zero.[1]

One of the nasty things about data in the social sciences is that although independent variables are usually correlated we are hardly ever sure that all the relevant variables were included. We can therefore rarely assume that the regression coefficients (b_j's) we obtain are unbiased estimators for the regression parameters (β_j's).

[1]It can also be shown that if X_j is not correlated with the remaining variables the b_i's for the remaining variables will be unaltered by the omission of X_j; for example, if $Y_R = 5.0 + 6.2X_1 + 6.3X_2 - 6.9X_3$ and X_3, a variable uncorrelated with X_1 or X_2, is dropped, the new regression will be $Y_R = C + 6.2X_1 + 6.3X_2$ (C is a constant).

▶ **Example 3.1** The data in Table 3.1 were run on the BMD03R multiple regression analysis program. The data values of Y were generated from

$$Y_R = -2 + 4X_1 + 4X_2 + 4X_3 + \epsilon$$

where $\sigma^2 = 0$; that is, the error term is zero.

In spite of some numerical stumbling,[1] the BMD03R program produced the expected results (Tables 3.2 and 3.3).

Table 3.1 Data for Example 3.1

X_1	X_2	X_3	Y
3	1	4	30
9	1	4	54
6	7	12	98
11	2	5	70
1	2	6	34
6	5	6	66

Table 3.2 Cross Products and Correlation Coefficients for Example 3.1

CROSS PRODUCTS OF DEVIATIONS

ROW 1	68.00000	.00000	− 5.00000	252.00000
ROW 2	.00000	30.00000	33.00000	252.00000
ROW 3	− 5.00000	33.00000	44.33333	291.33333
ROW 4	252.00000	252.00000	291.33333	3181.33333

CORRELATION COEFFICIENTS

ROW 1			
1.000	.2214E − 26	−.9056E − 01	.5418
ROW 2			
.2214E − 26	1.000	.8998	.8157
ROW 3			
−.9056E − 01	.8998	1.000	.7714
ROW 4			
.5418	.8157	.7714	1.000

[1]Although $\Sigma(X_{1i} - \bar{X}_1)(X_{2i} - \bar{X}_2)$, the cross product of deviations for X_1 and X_2 is equal to zero, $r_{X_1X_2}$ is given to be .2214E − 26 or $(.2214)10^{-26}$ when it also should be zero. The F value in Table 3.3 should be ∞; the asterisk in the printout indicates "overflow."

Table 3.3 BMD Regression Printout for Example 3.1

COEFFICIENT OF DETERMINATION	1.0000
MULTIPLE CORR COEFFICIENT	1.0000
SUM OF SQUARES ATTRIBUTABLE TO REGRESSION	3181.33333
SUM OF SQUARES OF DEVIATION FROM REGRESSION	0.00000
VARIANCE OF ESTIMATE	0.00000
STD ERROR OF ESTIMATE	0.00000
INTERCEPT (*A* VALUE)	− 2.00000

ANALYSIS OF VARIANCE FOR THE MULTIPLE LINEAR REGRESSION

SOURCE OF VARIATION	DF	SUM OF SQUARES	MEAN SQUARES	*F* VALUE
DUE TO REGRESSION	3	3181.33333	1060.44444	*93779.8442
DEVIATION ABT REGR	2	0.00000	0.00000	
TOTAL	5	3181.33333		

VARIABLE	MEAN	STD DEVIATION	REG COEFF	STD ERROR OF REG COEFF	COMPUTED *T* VALUE
1	6.00000	3.68782	4.00000	0.00000	0.00000
2	3.00000	2.44949	4.00000	0.00000	0.00000
3	6.16667	2.99444	4.00000	0.00000	0.00000
4(*Y*)	58.66667	25.22433			

TABLE OF RESIDUALS

OBS	*Y* VALUE	*Y* ESTIMATE	RESIDUAL
1	30.00000	30.00000	.00000
2	54.00000	54.00000	− 0.00000
3	98.00000	98.00000	− 0.00000
4	70.00000	70.00000	− 0.00000
5	34.00000	34.00000	− 0.00000
6	66.00000	66.00000	.00000

Suppose, however, that a researcher was unaware of the existence of independent variable X_3. He would run the regression of Y only on variables X_1 and X_2 and obtain the results in Table 3.4.

Table 3.4 Regression on the Data of Table 3.1 with X_3 Observations Omitted

COEFFICIENT OF DETERMINATION	.9589
MULTIPLE CORR COEFFICIENT	.9793
SUM OF SQUARES ATTRIBUTABLE TO REGRESSION	3050.68235
SUM OF SQUARES OF DEVIATION FROM REGRESSION	130.65098
VARIANCE OF ESTIMATE	43.55033
STD ERROR OF ESTIMATE	6.59927
INTERCEPT (*A* VALUE)	11.23137

ANALYSIS OF VARIANCE FOR THE MULTIPLE LINEAR REGRESSION

SOURCE OF VARIATION	DF	SUM OF SQUARES	MEAN SQUARES	*F* VALUE
DUE TO REGRESSION	2	3050.68235	1525.34118	35.0248
DEVIATION ABT REGR	3	130.65098	43.55033	
TOTAL	5	3181.33333		

VARIABLE	MEAN	STD DEVIATION	REG COEFF	STD ERROR OF REG COEFF	COMPUTED *T* VALUE
1	6.00000	3.68782	3.70588	.80028	4.63074
2	3.00000	2.44949	8.40000	1.20486	6.97179
3(*Y*)	58.66667	25.22433			

If someone now dared to claim that $\beta_2 = 4$, the researcher could run a hypothesis test:[1]

$$H_0 : \beta_2 = 4$$

$$H_1 : \beta_2 > 4$$

[1]If the researcher chose this one-tailed test merely because b_2 is greater than 4, he committed an error that is discussed in more detail in the next section.

Assuming the null hypothesis,

$$t(\text{for } b_2 = 8.4) = \frac{8.4 - 4}{1.20486} = 3.65^1$$

As seen in Figure 3.1, 3.65 falls in the rejection region of the t distribution with $n - m = 6 - 3 = 3$ degrees of freedom at the .05 level of significance.

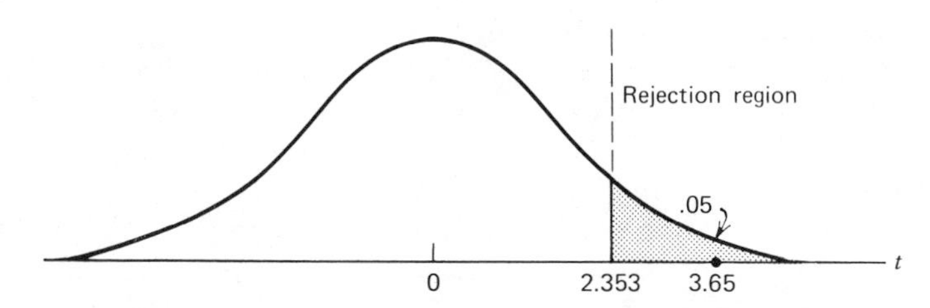

Figure 3.1 Rejection region for the t value of b_1.

We know, however, that $\beta_2 = 4.0$. What has occurred is evident from the correlation matrix in Table 3.2. Because X_2 and X_3 are highly correlated ($r_{X_2X_3} = .8998$), the estimator β_2 is biased because X_3 is missing. In fact, the estimate of 8.4 for the response of Y to X_2 includes the effect of X_3 on Y. Because X_1 and X_3 are also correlated, the estimator of β_1 is also biased. It is biased to a lesser degree because the correlation coefficient of X_1 and X_3 is small ($r_{X_1X_3} = -.09056$). ◀

In summary, multicollinearity is a condition that usually, but not necessarily, manifests itself in correlations among the independent variables in the data. A high degree of multicollinearity may lead to imprecision (large standard errors) in the estimation of regression parameters. The estimators, however, are still unbiased. In general, there is little that can be done about multicollinearity except to take a larger sample, preferably in a way that decreases multicollinearity.

When an independent variable that is correlated with others in the regression is not included and the regression parameter of this variable is not zero, the remaining coefficients will be biased estimators; but even if the omitted variable is not correlated (in the sample) with the remaining variables the estimators for the variances of the remaining coefficients $S_{b_i}^2$'s, will be biased upward (will tend to be too large).

[1]Actually, the t test is not appropriate. The "error" term is now due to a missing variable; it is not normally distributed and the original model breaks down. Further discussion of model breakdown is in Chapter 6.

This occurs because the "explanatory" power of the missing variable is removed, causing a larger sum of squared residuals, which, in turn, swells the variances of the regression coefficients. It becomes more difficult to show the significance of coefficients.

3.2 TESTS AND INTERPRETATIONS ON REGRESSION COEFFICIENTS

It is time to look somewhat more carefully at the partial regression coefficients. First we will examine some practical issues in hypothesis testing on coefficients. We will then return to the problem of estimation bias and its effect on imputations of causality and on attempts at using the regression for prediction.

A regression is often used as a test of some theory; for example, an economist may believe that the demand for a good decreases with an increase in the good's price (P_1) and increases with an increase in the price of a competing good (P_2). He would first check to see whether the signs of the regression coefficients are as expected; if so, they support the theory.

If a researcher believes that a parameter has a particular value, he can test this hypothesis against the alternative that the parameter does not have that value. Usually, however, the researcher will not be that definite in his preconceptions, hence will test the null hypothesis that the parameter is zero. In fact, many programs automatically print out t values that enable us to test this hypothesis for each parameter.

Two cautions are now in order. First, it is not proper to choose a one-tailed test *ex post* by noting whether the regression coefficient has a higher or lower value than expected. Second, one should not conclude that a variable is unimportant just because it fails a significance test against the hypothesis that the corresponding parameter is zero.

▶ **Example 3.2** Consider again the data of Table 2.1, which led to the computer printout in Table 2.2.

Suppose that our researcher (remember that we said he was naive) really had no idea how X_1, the number of stores in the downtown area, affects Y, the number of passengers per year.

Suppose further that he wishes to test the null hypothesis $H_0 : \beta_1 = 0$ at the .01 level of significance. He must assume $H_1 : \beta_1 \neq 0$. Table 2.2 gives $b_1 = 5.68$ and the corresponding t value is 2.92; $t_{.005}$ at $\nu = 20 - 7 = 13$ degrees of freedom is 3.012. As seen from Figure 3.2, 2.92 does not fall in the critical region and we cannot reject H_0; hence the regression coefficient is not significant.

We cannot impress someone with an "insignificant" coefficient. If, however, our researcher has a bit of guile, he could claim that he knew all

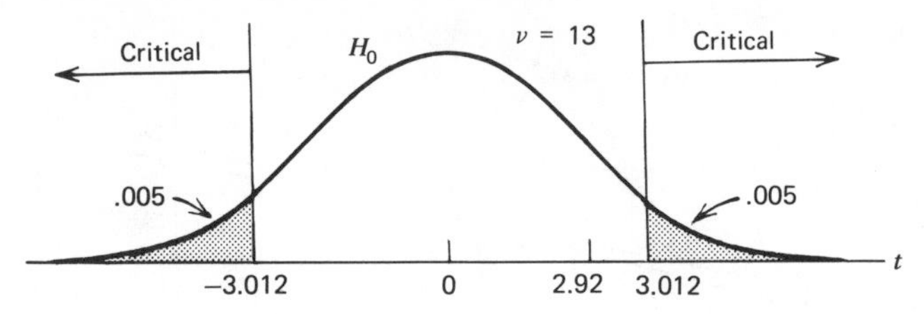

Figure 3.2 Rejection region for the t value of b_1 in Example 3.2.

along that stores attract people, and hence that β_1 is positive. The proper alternative hypothesis he could now claim is $H_1 : \beta_1 > 0$. The critical value of t is now $t_{.01} = 2.65$, and 2.92 abruptly becomes significant at the .01 level. A little bit of *ex post* theorizing can save the day. Unfortunately statisticians generally consider this sort of behavior unethical.

Lamentably, honesty can also lead our researcher into error. He may conclude that because b_1 is not significant at the .01 level the number of stores in the downtown area is not an important variable and he can regard β_1 as close to zero. Adoption of the rule that any t value greater than 3.012 or smaller than -3.012 produces rejection of the null hypothesis will indeed ensure that the probability of Type I error is less than .01. However, the probability of Type II error may be large. This means that the probability of ignoring an important parameter can be large.

Suppose that, unknown to the researcher, β_1 is actually 3. Let us assume that a coefficient of this size has important implications in his analysis of demand for transportation. What is the probability that this parameter will be dismissed as not significant? The probability of failing to reject the hypothesis $H_0 : \beta_1 = 0$ when actually $\beta_1 = 3.0$ can be calculated. This probability[1] is greater than .9. Of course, he could change the level of significance to .05 and obtain a somewhat smaller probability of Type II error.

The concluding moral in this example is that significance tests of $H_0 : \beta_i = 0$ are intended to guard against including a variable that is not important, a job they do well. However, we must ponder the consequences of neglecting a variable, especially if other evidence indicates that the significance test may have been failed largely because the sample was too small or the range in variation of the relevant independent variable was not large enough. ◀

Suppose that a linear relationship Y_R and an error term ϵ generated the Y values for the eight values of the independent variable X_1 in Figure 3.3. Suppose, further, that for some inscrutable reason only four points are to

[1]See Appendix A, Mathematical Note 3.1 for details of the calculation.

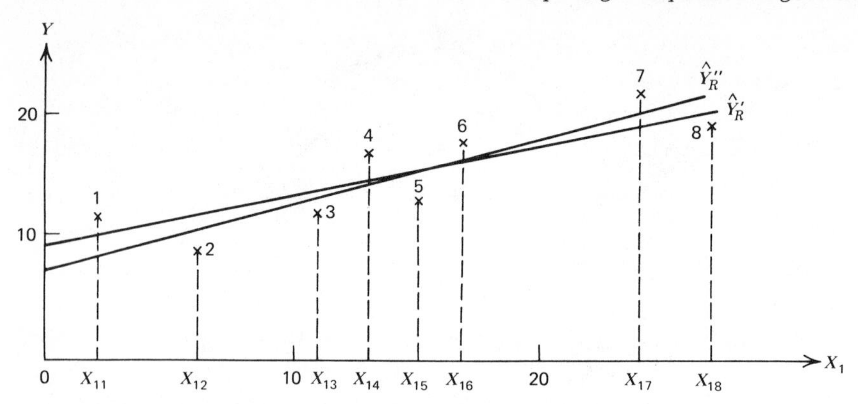

Figure 3.3 The effect of spread in the independent variable.

be chosen to find $\hat{Y}_R$. The choice of the points 1, 2, 7, and 8 leads to the line $\hat{Y}_R'$, and $S_{b_1} = .132$, whereas the choice of points 3, 4, 5, and 6 leads to the line $\hat{Y}_R''$ and $S_{b_1} = .606$. It is intuitively reasonable that there is more uncertainty in drawing the slope of a line or a plane through points closely grouped together than through points that are spread out. The conclusion is this: the spread of the independent variable in the sample may determine the estimated standard errors of the coefficients, hence affect the "significance" of the coefficients in hypothesis tests.

Let us now turn our attention to the problems in interpreting regression coefficients that may be caused by missing variables. In most observational, as opposed to experimental, studies, there will be missing variables; for example, in the study of Example 2.1 there may be many determinants of downtown transit demand that were not recorded by the researcher: weather conditions, ethnic makeup of the city, fare price history, available parking, the average number of muggings per bus ride, and so on.

As discussed in the last section, missing variables that are correlated with variables in the regression may cause bias in the regression coefficients. Unfortunately variables are often missing because they are unknown or because they cannot be practically measured. Therefore the extent of this bias is difficult to determine.

Sometimes the magnitude of the regression coefficients is important; this would be true in cases in which the independent variables are thought to have a causal effect on Y. Two remedies are occasionally feasible. We can include as many variables as possible as long as they have some effect on Y, even if they are not of direct interest or not controllable. We can also attempt to pick data in which any known missing variables are presumed to have approximately constant values. It can be shown that if the missing variables have small variances their biasing effect is small. Also, we should remember that even if an unbiased estimator for the parameter of a "causal" variable can be obtained it may not be that useful because it is often not possible to change values of that variable without also changing the values of other variables.

Finally, bias in regression coefficients may not be a disadvantage when the primary goal of the regression is prediction. If the correlation of the missing independent variables with the included independent variables is expected to continue, the bias in the coefficients will serve a purpose; as shown in the example in the last section it incorporates the prediction effect of the missing variables.

3.3 BETA COEFFICIENTS

We have not yet been able to compare the effectiveness of each of the independent variables $X_1, X_2, \ldots, X_q$ in predicting Y. The estimates $b_1, b_2, \ldots, b_q$ may not be used for such a ranking because they are dependent on the units of measurement of $X_1, X_2, \ldots, X_q$; for example, consider the regression line $\hat{Y}_R = 50 + 60X_1 + 370X_2$. If someone claimed that X_2 is more important then X_1, we could counter by defining new units for X_1 such that one new unit is 1000 times larger than the current unit (e.g., 180 units of Y are 60×3 grams or $60000 \times .003$ kilograms). The relationship now becomes $\hat{Y}_R = 50 + 60000X_1^* + 370X_2$, where X_1^* is in the new units.

A common way of obtaining coefficients that are comparable starts with dividing Y and $X_1, X_2, \ldots, X_q$ by their own standard deviations before regression begins; hence

$$Y_i' = \frac{Y_i}{S_Y}$$

$$X_{ji}' = \frac{X_{ji}}{S_{X_j}} \qquad j = 1, \ldots, q;\ i = 1, \ldots, n$$

If a regression is now calculated in the usual manner,

$$\hat{Y}_R' = \text{BETA}_0 + \text{BETA}_1 X_1' + \cdots + \text{BETA}_q X_q' \tag{3.6}$$

and the new parameters BETA_j are independent of the magnitude of units because each X_j' is a unitless number.[1] The *standardized regression*

[1]Sometimes Y_i' is defined as $(Y_i - \bar{Y})/S_Y$ and X_{ji}' as $(X_{ji} - \bar{X}_j)/S_{X_j}$. In this case $\text{BETA}_0 = 0$, but the other BETA's are the same as those in (3.6).

coefficient BETA_j can be interpreted as the number of standard deviations of Y that $\hat{Y}_R$ will change in response to an increase of one standard deviation in X_j (other independent variables being held constant).

It is not necessary to recalculate a regression to obtain the BETA_j's if the b_j's are known. We can use the conversion[2]

$$\text{BETA}_j = \frac{b_j S_{X_j}}{S_Y} \tag{3.7}$$

Many computer programs print the BETA coefficients as a matter of course.

▶ **Example 3.3** The BETA coefficients can be calculated from the information in Table 2.2. These calculations are summarized in Table 3.5. ◀

Table 3.5 BETA Coefficients Calculated from Table 2.2

$S_Y = 1.22$	j	S_{X_j}	b_j	BETA_j
	1	.158	5.68	.734
	2	.110	3.16	.285
	3	.117	.256	.0246
	4	.378	−.896	−.277
	5	.125	1.18	.120
	6	.0434	−1.88	−.0668

The usefulness of BETA coefficients should not be overstated. They are, of course, not immune to the problems caused by multicollinearity. Furthermore, the magnitude of the BETA coefficients are dependent on the variability of the independent variables in the particular sample that provided the data. Finally, independent variables may have economic or social importance (in addition to their predictive ability) because of the degree to which they are controllable.

3.4 INTERPRETING THE COEFFICIENT OF DETERMINATION

The coefficient of multiple determination is often erroneously regarded as a one-number summary measure of the quality of the regression obtained. We should remember, however, that R^2 may have high or low values as a result of chance or peculiarities in the data.

[2]See Appendix A, Note 3.2.

▶ **Example 3.4** A researcher obtains an R^2 value equal to .85. He fits an 18-parameter linear relationship with 20 data points. How impressive is his R^2? From (2.18)

$$F = \frac{20 - 18}{17} \cdot \frac{.85}{1 - .85}$$
$$= .667$$

Tables for F with $\nu_1 = 17$, $\nu_2 = 2$ show that the probability of an F higher than .667, hence of an R^2 greater than .85, occurring *purely by chance* (i.e., $\beta_1 = \beta_2 = \cdots = \beta_{17} = 0$) is[1] .75. Unfortunately researchers who do things like fitting 18-parameter relationships with 20 data points are not all that rare.[2] ◀

On the other hand, a low value of R^2 may argue not against the existence of a "good" linear relationship but may simply indicate that there was not enough variation in the Y values. Figure 3.3 gives an indication of how this might happen: $\hat{Y}_R''$ fitted through points 3, 4, 5, and 6 has $R^2 = .292$, whereas $\hat{Y}_R'$ fitted for points 1, 2, 7, and 8 was $R^2 = .830$. If the values of independent variables have little variation, thus causing the variance in Y_R values to be small in relation to the variance of the error term, we could obtain a low R^2.

A more insidious school of thought holds that the magnitude of R^2 is not so important as whether it (or, equivalently, F) is significant. Unfortunately, with large samples, even very small values of R^2 may be significant at, say, the .01 level of significance. This means that even though the fitted linear relationship $\hat{Y}_R$ has little explanatory value we may still be able to reject the hypothesis that all the slopes β_j are exactly 0.

▶ **Example 3.5** Suppose that a researcher has run a regression with 1000 data points. He has obtained a three-parameter ($m = 3$) linear relationship $\hat{Y}_R$ which has an R^2 of .01. Because

$$F = \frac{997}{2} \cdot \frac{.01}{.99} \qquad \text{(from (2.18))}$$
$$= 5.04$$

and because $F_{.01} \approx 4.62$ when $\nu_1 = 2$, $\nu_2 = 997$ ($F_{.01} = 4.62$ when $\nu_1 = 2$, $\nu_2 = 1000$) he could claim his R^2 is significant at the 1% level. ◀

[1]See Appendix A, Note 3.3.

[2]One rule of thumb is that 10 observations per variable are a minimum. Some thumbs, however, are thicker than others.

It is not entirely correct to say that a low R^2 means that a regression has no value. One or more of the coefficients may be significant and the corresponding parameter(s) may be of primary interest in the research. If, however, a large number of parameters is being estimated, we would expect some of them to appear significant by chance.[1]

A curious, but not uncommon, possibility is that R^2 is highly significant but that none of the regression coefficients is significantly different from zero. This may occur because multicollinearity increases the standard errors of coefficients. Dropping a variable or variables from the regression may appear to solve the problem by producing significance in some remaining variables. We must remember that the price to be paid may be bias in the remaining coefficients. The problem of deciding which variables to drop from the regression and which to retain is the main topic of the next chapter.

3.5 PREDICTION, INTERPOLATION AND EXTRAPOLATION

Although many researchers are concerned with the magnitudes of regression coefficients obtained, hoping to discover, measure, or test some causal relationship, others are interested mainly in using the linear relationship as a predictor. To make a point prediction[2] we need merely substitute in $\hat{Y}_R$ the values of $X_1, X_2, \ldots, X_q$ for which the prediction is to be made.

It should be noted, however, that reliance on extrapolation, that is, on predictions for values of $X_1, \ldots, X_q$ outside the region "covered" by the data, could be very dangerous. The relationship may actually be nonlinear (see Chapter 5), but our sample has picked a small "flat" point. Adding to the danger is the fact that in multiple regression it is difficult to tell if we are extrapolating; for example, consider a hypothetical researcher who has fitted the relationship $\hat{Y}_R = 5 + 7.3X_1 + 8.1X_2$. He wishes to make a prediction of Y for $X_1 = 5.5$ and $X_2 = 2.8$. He thinks that he is perfectly safe from extrapolation because the range of values for X_1 in his sample is [2, 6] and the range of values for X_2 is [2, 6].

[1]Suppose, for example, that there are 20 independent variables, no multicollinearity, and the values of Y were generated completely at random from a normal distribution. The probability that none of the coefficients will appear significantly different from zero at the .05 level of significance in t tests can be shown to be $(.95)^{20} = .358$. The probability of getting at least one spurious significance is therefore .642. This reminds us that the t test is valid for a *preselected* single coefficient.

[2]Using the notation developed in Appendix A, Note 2.1, we find that the estimated variance of this prediction can be shown as MSE $[1 + X_p'(X'X)^{-1}X_p]$, where X_p is derived from the vector of independent variables for which the prediction is being made and MSE is the mean squared error about regression.

Unfortunately, he is wrong. Figure 3.4 shows the projection of the data points on the X_1X_2 plane:

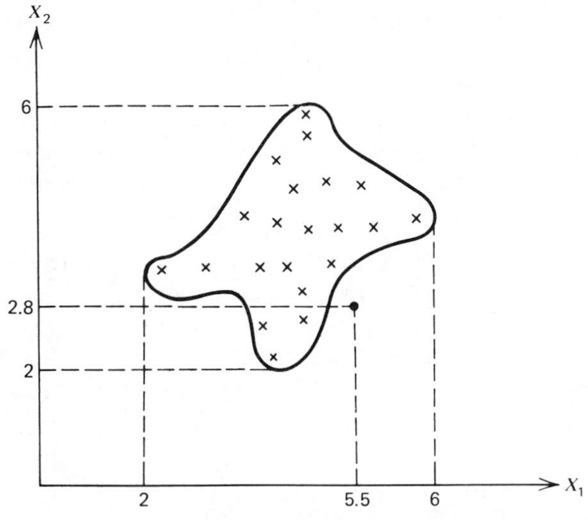

Figure 3.4 Extrapolation.

The enclosed area in Figure 3.4 could be the flat spot; the prediction required is outside its boundaries. With more than two independent variables extrapolation becomes even more difficult to detect.

EXERCISES

3.1 The following comments have been made about various regression studies. Discuss the truth, falsity, or incompleteness of each.

(a) Our initial computer run failed to show any significant regression coefficients. We removed the four variables with the lowest t values ($t = b_i / S_{b_i}$) and a new computer run established significance for the remaining three variables at the 1% level. There is no doubt that we have now isolated the three factors that govern land values.

(b) Multicollinearity can have no biasing effect on our parameter estimators because the highest intervariable value in the correlation matrix is .017.

(c) As long as you have a high R^2 you do not have to worry about the significance of the F value.

(d) We have significant regression coefficients and a significant F value and therefore do not have to worry about the fact that our R^2 is .001.

(e) It does not make any sense to test the hypothesis $H_0 : \beta_2 = 0$ against $H_1 : \beta_2 \neq 0$ because b_2 turned out to be large and positive.

(f) The t test for our hypothesis $H_0 : \beta_4 = 0$ gives a value of 1.93, whereas $t_{.025}$ is 2.045. I am afraid that this proves that advertising does not affect sales.

(g) I ran a regression with your variables and failed to get an R^2 significant at the 5% level. It is clear that your theory can do no better than chance in predicting inflation, hence is discredited forever.

3.2 Calculate the standardized regression (BETA) coefficients of the regression of Exercise 2.1 by using the appropriate standard deviations and the regular regression coefficients. Is it meaningful to rank the independent variables by BETA coefficients and to impute increasing importance to increasing magnitudes of BETA values? Discuss.

3.3 The geographer in Exercise 2.1 claimed that his regression formula would enable him to predict computer times for location problems with any number of facilities, potential sites, and demand centers. Discuss.

3.4 A colleague of the geographer in Exercise 2.1 claimed that he could have avoided multicollinearity by choosing from the data points a subset in such a way that correlations among X_1, X_2, and X_3 would have been minimized. How might this be done? Discuss the merits of the proposal.

REFERENCES

1. Jan Kmenta, *Elements of Econometrics*. New York: Macmillan, 1971.
2. Ralph E. Beals, *Statistics for Economists*. Chicago: Rand McNally, 1972.

chapter 4

Selecting variables for the regression equation

4.1 INTRODUCTION

Many researchers who use regression techniques have more independent variables than they wish to include in their final equation. This problem may arise because the researcher, while searching for a theory to call his own, is sifting through one of the proliferating data banks that computers have made possible; or he may honestly think that the dependent variable he is interested in has connections with many independent variables.

There are, of course, many good reasons for attempting a reduction in the number of independent variables. Aside from the principle of parsimony, there is no point in including variables that have little effect on the dependent variable. Then, too, it is often difficult or expensive to gather new information on many variables if validation or monitoring is needed. We usually need more than one computer run on any given data and computing time goes up sharply with the number of independent variables. Finally, researchers sometimes have a passion for isolating the "chief causes" of something and, of course, there should not be too many chief causes.

In view of the fact that there are now many easy-to-use computer programmed methods of selecting variables for the regression it is increasingly important to remember the price paid for parsimony. Part of the price, as discussed in Chapter 3, may be bias in the sample regression coefficients of the remaining variables and this in turn may be fatal to any attempt to measure the causative effect of an independent variable. Even if prediction is the primary motive we shall see that the coefficient of determination almost always decreases (and never increases) as variables are dropped.

There are many competing methods for selecting variables; each may produce a different result. Actually, even the criteria for reducing the number of independent variables are subject to dispute. The next two sections begin the discussion on criteria for adding or deleting variables.

4.2 PARTIAL F TESTS

Consider the case in which we have obtained a regression $\hat{Y}_R$ and in which

$$\hat{Y}_R = b_0 + b_1X_1 + b_2X_2 + \cdots + b_qX_q$$

Suppose that we wish to test the hypothesis $H_0 : \beta_j = 0$ against $H_1 : \beta_j \neq 0$ for some variable X_j. This could, of course, be done with a t test, discussed in Chapter 3. However, let us consider another statistic for testing H_0. Eventually this new statistic will enable us to test similar hypotheses regarding a number of parameters simultaneously.

First, let us define

$USS_{Y \cdot q}$ is the unexplained sum of squares of Y when $\hat{Y}_R$ has all q independent variables;

$USS_{Y \cdot q-1;j}$ is the unexplained sum of squares of Y when $\hat{Y}_R$ has only $q - 1$ independent variables because variable X_j has been removed.

According to Section 2.4, $USS_{Y \cdot q}$ has $(n - m)$ degrees of freedom and $USS_{Y \cdot q-1;j}$ has $n - (m - 1)$ degrees of freedom. It can be shown that $USS_{Y \cdot q-1;j} - USS_{Y \cdot q}$ has one degree of freedom (this seems reasonable because $n - (m - 1) - (n - m) = 1$).

The difference $USS_{Y \cdot q-1;j} - USS_{Y \cdot q}$ represents the improvement, in terms of removal of unexplained variation, obtained by using variable j in $\hat{Y}_R$. If this difference is large, we would expect to reject $H_0 : \beta_j = 0$. The difference is always nonnegative because the addition of a variable can never increase the unexplained sum of squares; the regression always has the option of setting the new coefficient to 0 in its "effort" to minimize the unexplained sum of squares.

It can be shown that the following ratio of mean squares

$$F = \frac{(USS_{Y \cdot q-1;j} - USS_{Y \cdot q})/1}{USS_{Y \cdot q}/(n - m)} \tag{4.1}$$

has an F distribution with $\nu_1 = 1$ and $\nu_2 = n - m$ whenever H_0 is true. The use of F is called a *partial F test* because it tests the hypothesis $H_0 : \beta_j = 0$ and not $H_0 : \beta_1 = \beta_2 = \cdots = \beta_q = 0$ as (2.15) does.

Also, it can be shown that[1]:

$$F = t^2 \tag{4.2}$$

where $t = (b_j - 0)/S_{b_j}$ in the regression equation in which variable X_j is included.

[1]Note the similarity with (1.10).

▶ **Example 4.1** Consider Example 3.1 again. After dropping variable X_3, the results of a regression in which the BMD03R program was used are as presented in Table 3.4. In particular,

$$\hat{Y}_R = 11.2 + \underset{(.80)}{3.71}\, X_1 + \underset{(1.2)}{8.40}\, X_2$$

Let us test the hypothesis $H_0 : \beta_2 = 0$ against $H_1 : \beta_2 \neq 0$.

$$t(\text{for } b_2 = 8.4) = \frac{8.4 - 0}{1.20}$$
$$= 6.97 \text{ (as printed out)}$$

We can reject the null hypothesis at the 0.01 level of significance because $t_{.005} = 5.841$ at $6 - 3 = 3$ degrees of freedom.

We can test the same hypothesis with the partial F test. To obtain $USS_{Y \cdot 1;\, 2}$ required by (4.1) we can use the results in Table 4.1, which is a BMD03R printout for a run with variable X_2 also omitted.

Table 4.1 Regression with X_2 Omitted

COEFFICIENT OF DETERMINATION	.2936
MULTIPLE CORR COEFFICIENT	.5418
VARIANCE OF ESTIMATE	561.86275
STD ERROR OF ESTIMATE	23.70364
INTERCEPT (*A* VALUE)	36.43137

ANALYSIS OF VARIANCE FOR THE MULTIPLE LINEAR REGRESSION

SOURCE OF VARIATION	DF	SUM OF SQUARES	MEAN SQUARES	*F* VALUE
DUE TO REGRESSION	1	933.88235	933.88235	1.6621
DEVIATION ABT REGR	4	2247.45098	561.86275	
TOTAL	5	3181.33333		

VAR	MEAN	STD DEV	REG COEFF	STD ERROR OF REG COEFF	COMPUTED *T* VALUE
1	6.00000	3.68782	3.70588	2.87449	1.28923
4(*Y*)	58.66667	25.22433			

In Table 3.4 $USS_{Y \cdot 2} = 130.65$ and in Table 4.1 $USS_{Y \cdot 1;\, 2} = 2247.45$. Hence

$$F = \frac{(2247.45 - 130.65)/1}{130.65/(6 - 3)}$$

$$= 48.6$$

Because $F_{.01} = 34.12$ at $\nu_1 = 1$ and $\nu_2 = 3$, we can reject H_0: $\beta_2 = 0$ at the .01 level of significance.

In passing, note that $t^2 = 6.97^2 \approx 48.6$ and that $t^2_{.005} = (5.841)^2 = 34.12 = F_{.01}$ in accordance with (4.2). ◀

The partial F test, although identical with the appropriate t test, is generally thought of in a different way. It is deemed a test of whether a variable in the regression has contributed significantly to reducing the unexplained variation in Y or, if the variable is not yet included in the regression, whether it would contribute significantly to reducing the unexplained variation in Y.

Its value also lies in the fact that it can be generalized to consider additions or deletions of two or more variables at a time. Let K be a set of k variables whose simultaneous equality to zero we wish to test. Suppose that we wished to test the hypothesis that $H_0 : \beta_j = 0, j \in K$ against the alternative hypothesis H_1 : (H_0 is not true). Then

$$F = \frac{(USS_{Y \cdot q-k;\, j \in K} - USS_{Y \cdot q})/k}{USS_{Y \cdot q}/(n - m)} \tag{4.3}$$

where $\nu_1 = k$ and $\nu_2 = n - m$. It is stressed that m is the number of parameters in the regression with K included. A demonstration of the use of this version of the partial F test is given in Example 5.5.

4.3 PARTIAL DETERMINATION AND CORRELATION

The *coefficient of partial determination* is that proportion of the unexplained sum of squares in Y that is removed by adding the independent variable.

$$r^2_{Yj \cdot q} = \frac{USS_{Y \cdot q-1;\, j} - USS_{Y \cdot q}}{USS_{Y \cdot q-1;\, j}} \tag{4.4}$$

The notation $r^2_{Yj \cdot q}$ is read as the partial determination of Y and variable X_j; after the addition of variable X_j there are q independent variables in the regression.[1] Other notation in (4.4) is that given in Section 4.2.

The similarity of this coefficient to the coefficient of determination should be apparent. Instead of measuring the proportion of the variation in Y explained by the total regression, the coefficient of partial determination gives that proportion of the variation "explained away" by adding independent variable X_j to the regression. The square root of the coefficient of partial determination is called the *partial correlation coefficient*.

The coefficient of partial determination is a number between zero and one. Coefficients near zero occur when $\text{USS}_{Y\cdot q-1;j} \approx \text{USS}_{Y\cdot q}$ and mean that variable X_j has provided little "explanation." If the coefficient is near one, (4.4) would lead us to conclude that $\text{USS}_{Y\cdot q}$ is nearly zero and that variable X_j has removed almost all the unexplained variation in Y.

It should be kept in mind that $r^2_{Yj\cdot q}$ is a statistic, hence has a probability distribution. In fact, it can be used as a test of the null hypothesis $H_0 : \beta_j = 0$. Its distribution is not usually tabled, but this presents no serious difficulties because it can be derived from the F or even the t tables.

It can be shown[2] that

$$F = \frac{(n - m)r^2_{Yj\cdot q}}{1 - r^2_{Yj\cdot q}} \tag{4.5}$$

where for F, $\nu_1 = 1$ and $\nu_2 = n - m$.

Note that $m = q + 1$ and is the number of parameters in the equation *after* the addition of variable X_j. Also

$$r^2_{Yj\cdot q} = \frac{F}{(n - m) + F} \tag{4.6}$$

A practical consequence of (4.2), (4.5), and (4.6) is that for any computer output that omits one or even two of partial F, t, or $r^2_{Yj\cdot q}$ the remaining statistics can easily be calculated.

▶ **Example 4.2** Let us turn to the basic problem of Example 4.1. Actually a part of the printout of the BMD03R program had previously been censored from Table 3.4. The truth can now be told and is presented in Table 4.2.

First let us check the partial correlation coefficient for variable X_2 by using (4.4), then (4.6).

[1]When the denominator $\text{USS}_{Y\cdot q-1;j}$ in (4.4) is replaced by $\text{USS}_{Y\cdot 0}$ (this is the total SS as in (2.14)), the resulting coefficient is called *the coefficient of part determination*. Its square root is the *coefficient of part correlation*. The reader should ponder on the difference between part and partial coefficients.

[2]Appendix A, Note 4.1.

Table 4.2 Addition to Table 3.4

VARIABLE	PARTIAL CORR COEFF	SUM OF SQUARES ADDED	PROP VAR CUM
1	.93663	933.88235	.29355
2	.97050	2116.80000	.66538

As found in Example 4.1, $USS_{Y \cdot 2} = 130.65$ and $USS_{Y \cdot 1;\, 2} = 2247.45$. From (4.4)

$$r^2_{Y2 \cdot 2} = \frac{2247.45 - 130.65}{2247.45}$$

$$= .942$$

Note that the difference $2247.45 - 130.65 = 2116.8$ was printed out by BMD03R under the "sum of squares added" column. The heading reflects the interpretation that the sum of squares removed from the unexplained category must be added to the sum of squares explained by the regression. The printout, however, does not provide information for calculating the numerator of $r^2_{Y1.2}$ in the same way; the sum of squares 933.88235 is added with variable X_2 absent, not held constant. The "proportion of variance cumulative" column gives the accumulated explained sum of squares as a proportion of the total sum of squares (in Table 3.4).

Alternatively, from (4.6) and the partial F value of 48.6 found in Example 4.1,

$$r^2_{Y2 \cdot 2} = \frac{48.6}{(6 - 3) + 48.6}$$

$$= .942$$

The partial correlation coefficient is $\sqrt{.942} = .971$ as in Table 4.2.

Suppose that we wish to test $H_0 : \beta_2 = 0$ against $H_1 : \beta_2 \neq 0$ with $r^2_{Y2 \cdot 2}$. The critical value calculated from (4.6) with $F_{.01} = 34.12$ is

$$r^2_{.01} = \frac{34.12}{(6 - 3) + 34.12}$$

$$= .920$$

Again, we reject H_0 at the .01 level of significance. ◀

4.4 FORWARD SELECTION, BACKWARD ELIMINATION, AND STEPWISE REGRESSION

In this section three techniques for selecting variables for a regression are discussed. An evaluation and comparison of other methods is given in [1].

All three methods are currently popular and many computer programs exist for them. This section presents a straightforward and rather uncritical description of these procedures; the next devotes itself to "flies in the ointment."

In the *forward selection* procedure we start with no variables in the regression equation. The first independent variable considered for insertion into the equation has the highest coefficient of partial determination.[1] Note that the variable with the highest coefficient of partial determination will also have the highest partial F value and would have the highest absolute value of t (as given by b_0/S_{b_0}) if it were entered in the equation.[2] Therefore, if a computer program prints out partial F or t values for the "candidate" variables, they can be used alternatively as a criterion.

It is usual to set a constant value F_{in} such that the partial F value of the best candidate variable must be equal to or greater than F_{in} if that variable is to be entered. Forward selection proceeds until no candidate variables qualify or all the independent variables have entered the equation.

Backward elimination is the opposite of the forward selection procedure. All the independent variables start in the equation and are eliminated one at a time. At each stage the variable in the equation that has the lowest coefficient of partial determination (or the lowest partial F or t values) is eliminated, provided that its partial F is less than some constant F_{out}.

The *stepwise regression* procedure is an improved version of forward selection. In this procedure the variables already in the equation are reevaluated at each stage. Because of intercorrelation, a variable that was important at an earlier stage may not be important at a later one. In stepwise regression, before a variable is added, the variable already in the regression with the lowest partial F value is dropped if this value is less than F_{out}. This procedure is summarized in the flow diagram of Figure 4.1. Note that F_{in} must be greater than or equal to F_{out} or an infinite loop could result.

In the following two examples the data of Example 2.1 are run first on a stepwise regression, then on a backward elimination procedure.

▶ **Example 4.3** The data in Table 2.1 were run on the SPSS [2] stepwise program. F_{in} and F_{out} were both arbitrarily set at 1.0.

For the first step the variable that would produce the highest partial F value is entered (note that with only one variable in $\hat{Y}_R$ the F value *is* the partial F value). Actually, this variable could have been selected by

[1]To calculate the coefficient of partial determination from (4.4) when a variable is the first to be entered, note that $USS_{Y\cdot 0;\,j}$ is equal to the total sum of squares or $\Sigma(Y_i - \bar{Y})^2$. The reader should also convince him/herself [by pondering on (2.16) and (4.4)] that the variable with the highest coefficient of partial determination will cause the biggest increase in R^2 on addition.

[2]By (4.2) the partial F increases with t^2 and vice versa, and by (4.5) the partial F increases with $r^2_{Yj\cdot q}$.

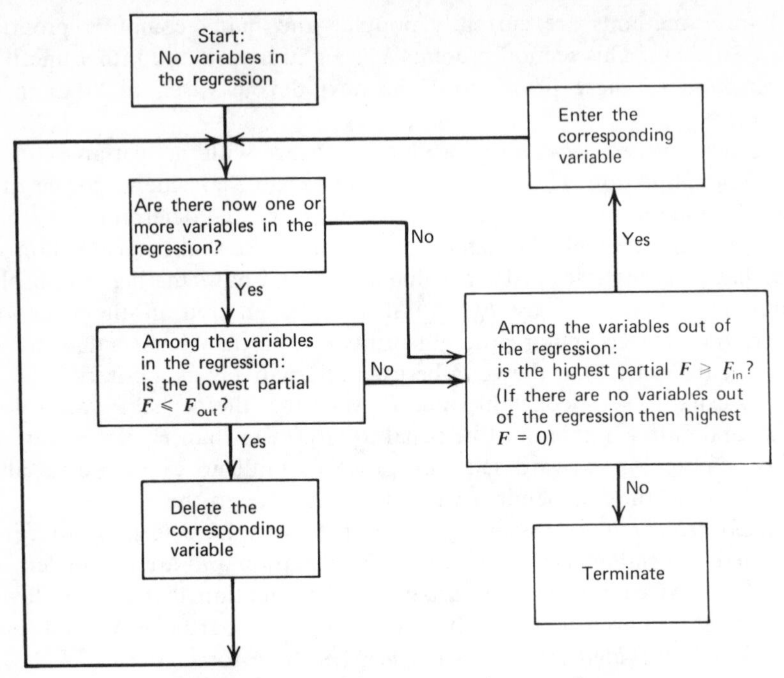

Figure 4.1 The stepwise regression procedure.

choosing that independent variable with the highest correlation to the dependent variable Y in Table 4.3 because F and R^2 increase together as is evident from (2.18). Because the F value of X_1 is 53.95093 and is higher than 1.0, X_1 is entered in the regression in Table 4.4.

For a step in the stepwise procedure the partial F values of variables already in the regression are checked first. In this case X_1, the only variable in the regression, has a partial F greater than F_{out}. Of the variables not in

Table 4.3 Correlation Coefficients for the Data in Table 2.1

CORRELATION COEFFICIENTS

X_2	.10358					
X_3	.04585	−.15896				
X_4	.16390	−.19067	.28347			
X_5	.75200	.13540	−.00943	.48790		
X_6	−.84018	.22498	.13757	−.21853	−.64902	
Y	.86593	.41077	−.07583	−.13082	.61870	−.63373
	X_1	X_2	X_3	X_4	X_5	X_6

Table 4.4 Variable X_1 Is in the Regression Equation

VARIABLE(S) ENTERED ON STEP NUMBER 1 . . . X_1

		F	SIGNIFICANCE[1]
MULTIPLE R	.86593		
R SQUARE	.74983	53.95093	0

ANALYSIS OF VARIANCE	DF	SUM OF SQUARES	MEAN SQUARE
REGRESSION	1	21.30317	21.30317
RESIDUAL	18	7.10751	.39486

VARIABLES IN THE EQUATION

VAR	B	STD ERROR B	F SIGNIFICANCE[1]	BETA ELASTICITY[2]
X_1	6.6960768	.91163489	53.950935	.8659270
			.000	.57520
CONST	4.6881192	.87557773	28.668692	
			.000	

VARIABLES NOT IN THE EQUATION

VARIABLE	PARTIAL[3]	TOLERANCE[4]	F SIGNIFICANCE[1]
X_2	.64540	.98927	12.136338
			.003
X_3	−.23123	.99790	.96024915
			.341
X_4	−.55277	.97314	7.4800422
			.041
X_5	−.09850	.43450	.16656569
			.688
X_6	.34581	.29410	2.3090913
			.147

[1]The SIGNIFICANCE in this program is Pr($F >$ the printed F value) e.g., Pr($F > 53.95$) = .000. If the significance level for variable X_j is, say, .01, then a SIGNIFICANCE of less than or equal to .01 means rejection of the null hypothesis that $\beta_j = 0$. Consequently SPSS eliminates the need to look up F tables.

[2]The elasticity is roughly a measure of the percent change in Y caused by a 1% change in X_j. In this program ELASTICITY = $b_j \cdot \bar{X}_j / \bar{Y}$. See Appendix A, Note 4.2.

[3]This is the partial correlation coefficient ($r_{Yj \cdot q}$).

[4]The TOLERANCE is a number between 0 and 1. If the tolerance is near 0, then the variable is close to being a linear combination of the variables already in the equation. Variables with low tolerances could cause computational difficulties in the matrix method of calculation. A minimum tolerance for the entry of a variable may be set by the program user or by the program itself.

the equation, X_2 has the highest partial F value;[1] this value is greater than F_{in} and X_2 enters in Table 4.5

All the variables in the regression at the end of Step 2 have partial F levels greater than 1; hence none is deleted. Of the variables not in the regression X_4 has the highest partial F level and, in addition, $7.215 > 1$. Variable X_4 is entered with the result of Table 4.6.

Table 4.5 Variables X_1 and X_2 Are in the Regression Equation

VARIABLE(S) ENTERED ON STEP NUMBER 2 . . . X_2

			F	SIGNIFICANCE
MULTIPLE R	.92414			
R SQUARE	.85403		49.73296	0
STD DEVIATION	.49390			

ANALYSIS OF VARIANCE	DF	SUM OF SQUARES	MEAN SQUARE
REGRESSION	2	24.26370	12.13185
RESIDUAL	17	4.14698	.24394

VARIABLES IN THE EQUATION

VAR	B	STD ERROR B	F / SIGNIFICANCE	BETA / ELASTICITY
X_1	6.4361129	.72041401	79.814721	.8323088
			0	.55287
X_2	3.6073028	1.0354730	12.136338	.3245539
			.003	.32066
(CONST)	1.3958010	1.1690805	1.4254701	
			.249	

VARIABLES NOT IN THE EQUATION

VARIABLE	PARTIAL	TOLERANCE	F / SIGNIFICANCE
X_3	−.16576	.97081	.45206144
			.511
X_4	−.55748	.92955	7.2149468
			.016
X_5	−.20385	.43115	.69369764
			.417
X_6	−.04416	.19569	3.12646079E-02
			.862

[1] Variables with the highest (lowest) partial F values will, of course, also have the highest (lowest) absolute values of the partial correlation coefficient.

Table 4.6 Final Regression Equation

VARIABLE(S) ENTERED ON STEP NUMBER 3 . . . X_4

		F	SIGNIFICANCE
MULTIPLE R	.94837		
R SQUARE	.89940	47.68138	0
STD DEVIATION	.42265		

ANALYSIS OF VARIANCE	DF	SUM OF SQUARES	MEAN SQUARE
REGRESSION	3	25.55254	8.51751
RESIDUAL	16	2.85814	.17863

VARIABLES IN THE EQUATION

VAR	B	STD ERROR B	F / SIGNIFICANCE	BETA / ELASTICITY
X_1	6.7532423	.62768899	115.75390	.8733195
			.000	.58011
X_2	3.0919349	.90662789	11.630611	.2781855
			.004	.27484
X_4	−.71443791	.26597935	7.2149468	−.2209126
			.016	−.07833
(CONST)	2.4652080	1.0767364	5.2418958	
			.036	

VARIABLES NOT IN THE EQUATION

VAR	PARTIAL	TOLERANCE	F / SIGNIFICANCE
X_3	−.02987	.90802	1.33989057E-02
			.909
X_5	.19168	.27848	.57215639
			.461
X_6	−.07739	.19544	9.03787809E-02
			.768

F LEVEL OR TOLERANCE LEVEL INSUFFICIENT FOR FURTHER COMPUTATION

In Table 4.6 no variables qualify for leaving or entry and the procedure terminates. Note that because no variables were ever deleted after entry the forward selection procedure would have given the identical results. ◀

▶ **Example 4.4** The data in the preceding example were run on the SPSS backward elimination procedure. The same levels of F_{out} and tolerance were used: 1.0 and .001, respectively.

In Step 1 all variables were entered (Table 4.7). The partial F value of X_3 was lowest and less than 1.0; hence X_3 was deleted.

Table 4.7 All Variables Are in the Regression Equation

VARIABLE(S) ENTERED ON STEP NUMBER 1 . . . X_1 X_3 X_2 X_4 X_5 X_6

MULTIPLE R	.95066	F	SIGNIFICANCE
R SQUARE	.90375	20.34381	0
STD DEVIATION	.45864		

ANALYSIS OF VARIANCE	DF	SUM OF SQUARES	MEAN SQUARE
REGRESSION	6	25.67611	4.27935
RESIDUAL	13	2.73457	.21035

VARIABLES IN THE EQUATION

VAR	B	STD ERROR B	F / SIGNIFICANCE	BETA / ELASTICITY
X_1	5.6794079	1.9457961	8.5194483	.7344528
			.012	.48787
X_3	.25607531	1.1594154	4.87817442E-02	.0245667
			.829	.16229
X_2	3.1629382	1.3431774	5.5451671	.2845738
			.035	.28116
X_4	−.89579207	.39040218	5.2648981	−.2769895
			.039	−.09822
X_5	1.1765064	1.6589505	.50294643	.1198961
			.491	.10250
X_6	− 1.8798275	6.5623233	8.20579663E-02	−.0667809
			.779	−.05255
(CONST)	1.2907125	7.1883193	3.22406899E-02	
			.860	

After Step 2 (Table 4.8) X_6 qualifies for deletion.

Table 4.8 Regression Equation with X_3 Deleted

VARIABLE(S) REMOVED ON STEP NUMBER 2 . . . X_3

MULTIPLE R	.95047	F	SIGNIFICANCE
R SQUARE	.90339	26.18172	0
STD DEVIATION	.44279		

ANALYSIS OF VARIANCE	DF	SUM OF SQUARES	MEAN SQUARE
REGRESSION	5	25.66585	5.13317
RESIDUAL	14	2.74483	.19606

Table 4.8 (continued)

VARIABLES IN THE EQUATION

VAR	B	STD ERROR B	F SIGNIFICANCE	BETA ELASTICITY
X_1	5.9147633	1.5717882	14.160764 .002	.7648886 .50808
X_2	3.0501442	1.1993589	6.4675894 .023	.2744256 .27113
X_4	−.86217521	.34708202	6.1705920 .026	−.2665947 −.09453
X_5	1.0905286	1.5568836	.49063871 .495	.1111342 .09501
X_6	− 1.0967400	5.3311760	4.23215274E-02 .840	−.0389617 −.03066
(CONST)	2.7696460	2.5239276	1.2041893 .291	

VARIABLES NOT IN THE EQUATION

VAR	PARTIAL	TOLERANCE	F SIGNIFICANCE
X_3	.06114	.59845	4.87817442E-02 .829

After Step 3 (Table 4.9) X_5 has a partial F lower than 1.0 and is removed.

Table 4.9 Regression Equation with X_3 and X_6 Deleted

VARIABLE(S) REMOVED ON STEP NUMBER 3 . . X_6

		F	SIGNIFICANCE
MULTIPLE R	.95031		
R SQUARE	.90310	34.94782	0
STD DEVIATION	.42842		

ANALYSIS OF VARIANCE	DF	SUM OF SQUARES	MEAN SQUARE
REGRESSION	4	25.65755	6.41439
RESIDUAL	15	2.75313	.18354

Table 4.9 (continued)

VARIABLES IN THE EQUATION

VAR	B	STD ERROR B	F / SIGNIFICANCE	BETA / ELASTICITY
X_1	6.1556807	1.0143536	36.827597	.7960437
			.000	.52878
X_2	2.9805784	.95043006	9.3652895	.2616887
			.008	.25855
X_4	−.86543725	.33546889	6.6552852	− 0.2676034
			.021	−.09489
X_5	1.1305146	1.4945787	.57215639	.1152091
			.461	.09849
(CONST)	2.3072887	1.1112162	4.3112851	
			.055	

VARIABLES NOT IN THE EQUATION

VARIABLE	PARTIAL	TOLERANCE	F / SIGNIFICANCE
X_3	.02171	.84516	6.60272761E-03
			.936
X_6	−.05490	.19239	4.23215274E-02
			.840

Step 4 (Table 4.10) is the last step.

Table 4.10 Regression Equation with X_3, X_5, and X_6 Deleted

VARIABLE(S) REMOVED ON STEP NUMBER 4 . . . X_5

MULTIPLE R	.94837	F	SIGNIFICANCE
R SQUARE	.89940	47.68138	0
STD DEVIATION	.42265		

ANALYSIS OF VARIANCE	DF	SUM OF SQUARES	MEAN SQUARE
REGRESSION	3	25.55254	8.51751
RESIDUAL	16	2.85814	.17863

Table 4.10 (continued)

VARIABLES IN THE EQUATION

VAR	B	STD ERROR B	F / SIGNIFICANCE	BETA / ELASTICITY
X_1	6.7532423	.62768899	115.75390	.8733195
			.000	.58011
X_2	3.0919349	.90662789	11.630611	.2781855
			.004	.27484
X_4	−.71443791	.26597935	7.2149468	−.2209126
			.016	−.07833
(CONST)	2.4652080	1.0767364	5.2418958	
			.036	

VARIABLES NOT IN THE EQUATION

VAR	PARTIAL	TOLERANCE	F / SIGNIFICANCE
X_3	−.02987	.90802	1.33989057E-02
			.909
X_5	.19168	.27848	.57215639
			.461
X_6	−.07739	.19544	9.03787809E-02
			.768

F LEVEL OR TOLERANCE LEVEL INSUFFICIENT FOR FURTHER COMPUTATION

Comparison of the stepwise and backward elimination procedures shows that in this case they arrived at the same solution. The SPSS printout omits one statistic of interest, the adjusted coefficient of determination $\overline{R}^2$ [see (2.17)]. It would be interesting to see what happens to $\overline{R}^2$ in the backward elimination procedure.

For Step 1, using Table 4.7,

$$\overline{R}^2 = 1 - \frac{\text{mean square about regression (residual)}}{\text{mean square about the mean}}$$

$$= 1 - \frac{0.2104}{28.41/19}$$

$$= .860$$

Table 4.11 gives $\overline{R}^2$ for Steps 1, 2, 3, and 4.

Table 4.11 Values of $\bar{R}^2$ in the Backward Elimination

Step	$\bar{R}^2$
1	.860
2	.869
3	.877
4	.881

We note that although R^2 decreased in Steps 1 through 4, $\bar{R}^2$ actually increased. More is said about this phenomenon in the next section. ◀

In this section the rules for forward selection, backward elimination, and stepwise regression procedures were given. Many questions should still remain. Section 4.5 discusses such issues as the choice of F values for inclusion and deletion, bias in coefficients, and the imposition of additional rules for inclusion and deletion.

4.5 DESIGNING AND INTERPRETING SELECTION PROCEDURES

The selection procedures discussed in Section 4.4 seem eminently reasonable in principle. Variables are chosen to enter or leave on the basis of the highest or lowest partial F value. Because this is equivalent to using the highest or lowest coefficient of partial determination, we are choosing entrance for those variables with the most power to dispel unexplained variation in the dependent variable and deletion for those with the least.

The first practical difficulty is the choice of the F threshold values. We could, for example, try to set F_{in} in the stepwise procedure at $F_{.01}$, the 1% significance level. However, because the degrees of freedom change step by step, so does $F_{.01}$. This is not a serious problem if n, the number of data points, is much greater than m, the number of parameters being fitted, because $F_{.01}$ will remain fairly stable [see (4.1) and Table 3 in Appendix C].

Of more importance is the choice of the level of significance to approximate. Using $F_{\text{in}} \approx F_{.01}$ means that we are allowing one chance in 100 for admitting a variable X_j when β_j is actually 0. Thus we are relatively safe from Type I error, but at a price. The probability of not including the variable X_j when β_j is large enough to be important in the study may be quite large (see Section 3.2). Similar problems of balancing Type I and Type II errors exist in choosing F_{out}.

Compounding the difficulty may be the researcher's preferences for including certain variables or excluding others. Although selection programs usually do not provide for individual F_{in} and F_{out} values for different variables, they may cater to prejudice in a different way. Priority levels may be assigned to variables so that all variables with a higher priority level will enter first. The basic selection procedures so modified have great flexibility. An advantage is that the researcher may apply his theoretical knowledge of the problem in conjunction with statistical methods. On the other hand, this flexibility, when used to "play" with the regression, may accomplish nothing more than a large bill for computer time.

The forward selection procedure adds the variable that maximizes the increase in R^2 at each stage. This does not mean, however, that it will always find the four variables, for example, which when chosen from the six available independent variables will give the largest possible R^2. (It always would if the six variables were uncorrelated.) The backward elimination procedure drops that variable at each stage that contributes least to R^2, but it, too, does not guarantee to find the equation with maximum R^2 for a given number of variables. If a researcher had enough computer time, he could use the "*all possible regressions procedure;*" that is, he could, for any given number of independent variables, try all possible ways of choosing that number and record R^2 for each combination. The forward selection and backward elimination procedures have the virtue of being faster.

Suppose that the researcher has $\bar{R}^2$, the adjusted coefficient of determination, as a criterion. It can be shown [3] that if the partial F of a variable not in the equation is greater than 1.0 then $\bar{R}^2$ will be increased if that variable is added. Similarly, if the partial F of a variable in the equation is less than 1.0, dropping that variable will again increase $\bar{R}^2$. Choosing $F_{in} = F_{out} = 1.0$ in stepwise regression is therefore consistent with maximizing $\bar{R}^2$, but does not guarantee that a maximum $\bar{R}^2$ will be achieved [4].

The design, as well as the interpretation, of selection procedures is further complicated by our old nemesis, multicollinearity. Suppose that (although it is unlikely) all relevant independent variables are initially available. After selection the sample regression coefficients may be quite biased if multicollinearity was serious (one advantage of the backward elimination procedure over the stepwise procedure is that changes in coefficients can be monitored by comparison with the "all included" state), in which case we must often give up any attempt to interpret coefficients.

Other disturbing phenomena may occur: different procedures may give widely different results. Thus we may have a choice of regressions with perhaps roughly the same number of variables and virtually the same R^2 and $\bar{R}^2$ but with a different selection of variables and different coefficients for those independent variables that are in common. Sometimes variables that are the first to enter the equation may not be included in the final regression.

▶ **Example 4.5** Consider the set of data points in Table 4.12.[1]

Table 4.12 Data for Example 4.5

X_1	X_2	X_3	X_4	X_5	X_6	Y
40	24	2000	3	0	1	385
12	17	1000	6	5	5	158
20	18	2200	5	4	4	201
32	12	1500	4	7	4	238
16	21	6050	0	2	6	101
35	25	2100	7	1	1	397
21	16	3000	2	7	5	140
17	21	2100	1	1	5	155
14	12	1500	1	8	6	90
30	18	2800	3	5	4	342
25	16	1200	8	7	3	257
28	20	4000	12	3	2	356
35	16	5500	10	8	3	324

A regression on all six independent variables was run on the SPSS program; the results are presented in Tables 4.13 and 4.14.

Table 4.13 Means, Standard Deviations, and Correlation Coefficients for the Data in Table 4.12

VARIABLE	MEAN	STD DEVIATION	CASES
X_1	25.0000	9.0554	13
X_2	18.1538	3.9968	13
X_3	2688.4615	1589.8315	13
X_4	4.7692	3.6777	13
X_5	4.4615	2.8465	13
X_6	3.7692	1.6909	13
Y	241.8462	109.5938	13

CORRELATION COEFFICIENTS

X_2	.32235					
X_3	.10853	.20161				
X_4	.41037	.04797	.10854			
X_5	−.17458	−.92236	−.07238	.14635		
X_6	−.84905	−.54921	.07023	−.62572	.38757	
Y	.89605	.47891	.03440	.61272	−.30562	−.93469
	X_1	X_2	X_3	X_4	X_5	X_6

[1]Reprinted by special permission from *Statistics: A Foundation for Analysis* by Anne Hughes and Dennis Gravoig, Addison-Wesley Publishing Company, Inc., Copyright © 1971.

Table 4.14 SPSS Regression Printout for Example 4.5

MULTIPLE *R*	.97544		*F*	SIGNIFICANCE
R SQUARE	.95149		19.61326	0
STD DEVIATION	34.13719			
			SUM OF	MEAN
ANALYSIS OF VARIANCE		DF	SQUARES	SQUARE
REGRESSION		6	137137.60622	22856.26770
RESIDUAL		6	6992.08609	1165.34768

VAR	*B*	STD ERROR *B*	*F* / SIGNIFICANCE	BETA / ELASTICITY
X_1	12.781241	4.6426493	7.5790422	1.0560728
			.033	1.32122
X_3	− 2.03283687E-02	1.29242876E-02	2.4739552	−.2948951
			.167	−.22598
X_5	1.4502648	10.926305	1.76176501E-02	.0376680
			.899	.02675
X_4	17.076585	7.9950981	4.5619914	.5730518
			.077	.33675
X_2	15.100350	10.550742	2.0483682	.5506969
			.202	1.13349
X_6	40.774867	40.589186	1.0091702	.6290883
			.354	.63549
(CONST)	− 538.76475	432.53483	1.5515157	
			.259	

From Table 4.14

$$\hat{Y}_R = -539 + \underset{(4.64)}{12.8}\,X_1 + \underset{(10.6)}{15.1}\,X_2 - \underset{(.0129)}{.0203}\,X_3$$

$$+ \underset{(8.00)}{17.1}\,X_4 + \underset{(10.9)}{1.45}\,X_5 + \underset{(40.6)}{40.8}\,X_6$$

where the numbers in brackets are standard errors of the coefficients. Accordingly, from the corresponding partial *F*'s and their SIGNIFICANCE printouts, only b_1 is significant at the .05 level of significance and no coefficients are significant at the .01 level. According to the correlation matrix, however, considerable correlation is present among the independent variables; hence multicollinearity adds to the standard errors of the coefficients.

Stepwise regression and backward elimination procedures were applied to the problem by using $F_{in} = F_{out} = 1.0$ for the first and $F_{out} = 1.0$ for the second. The results are presented in summary form in Tables 4.15 and 4.16.

Table 4.15 Summary Table: Stepwise Procedure with $F_{in} = F_{out} = 1$

	VARIABLE								
STEP	ENTD	REMD	F TO ENTER OR REMOVE	SIGNIF	MULT R	R SQ	R SQUARE CHANGE	OVERALL F	SIGNIFICANCE
1	X_6		76.06189	.000	.93469	.87365	.87365	76.06189	0
2	X_1		4.23722	.067	.95460	.91126	.03760	51.34180	0
3	X_4		1.06421	.329	.95950	.92064	.00938	34.80237	0
4	X_2		1.24754	.296	.96506	.93135	.01071	27.13156	0
5		X_6	.49154	.503	.96287	.92713	−.00422	38.16789	0
6	X_3		2.26786	.171	.97120	.94322	.01610	33.22552	0
7	X_6		1.16853	.316	.97537	.95135	.00812	27.37408	0

Table 4.16 Summary Table: Backward Procedure with $F_{out} = 1$

	VARIABLE								
STEP	ENTD	REMD	F TO ENTER OR REMOVE	SIGNIF	MULT R	R SQ	R SQUARE CHANGE	OVERALL F	SIGNIFICANCE
1	X_1		7.57904	.033	.89605	.80290	.80290	19.61326	0
	X_3		2.47396	.167	.89827	.80690	.00400		
	X_5		.01762	.899	.91158	.83098	.02409		
	X_4		4.56199	.077	.96688	.93485	.10387		
	X_2		2.04837	.202	.97125	.94333	.00848		
	X_6		1.00917	.354	.97544	.95149	.00816		
2		X_5	.01762	.899	.97537	.95135	−.00014	27.37408	0

Table 4.17 Summary Table: Stepwise Procedure with $F_{in} = F_{out} = 1.2$

STEP	VARIABLE ENTD	VARIABLE REMD	F TO ENTER OR REMOVE	SIGNIF	MULT R	R SQ	R SQUARE CHANGE	OVERALL F	SIGNIF
1	X_6		76.06189	.000	.93469	.87365	.87365	76.06189	0
2	X_1		4.23722	.067	.95460	.91126	.03760	51.34180	0

Table 4.18 Summary Table: Backward Procedure with $F_{out} = 1.2$

STEP	VARIABLE ENTD	VARIABLE REMD	F TO ENTER OR REMOVE	SIGNIF	MULT R	R SQ	R SQUARE CHANGE	OVERALL F	SIGNIF
1	X_1		7.57904	.033	.89605	.80290	.80290	19.61326	0
	X_3		2.47396	.167	.89827	.80690	.00400		
	X_5		.01762	.899	.91158	.83098	.02409		
	X_4		4.56199	.077	.96688	.93485	.10387		
	X_2		2.04837	.202	.97125	.94333	.00848		
	X_6		1.00917	.354	.97544	.95149	.00816		
2		X_5	.01762	.899	.97537	.95135	−.00014	27.37308	0
3		X_6	1.16853	.316	.97120	.94322	−.00812	33.22552	0

The final equation in both cases is

$$\hat{Y}_R = -515 + \underset{(4.30)}{12.8}\,X_1 + \underset{(6.71)}{14.1}\,X_2 - \underset{(.0118)}{.0200}\,X_3 + \underset{(7.33)}{17.2}\,X_4 + \underset{(37.6)}{40.7}\,X_6$$

Only the coefficient of X_1 is significant at the .05 level.

Some researchers have a tendency to judge the "worth" of an independent variable by the change that it causes in the coefficient of determination. As Table 4.15 shows, the change in R^2 can be an unstable measure if multicollinearity is present. When X_6 first enters, $\Delta(R^2)$ is .87365 (this is also the square of the simple correlation coefficient of Y and X_6). After it has been deleted and reenters its $\Delta(R^2)$ is .00812. If there were no correlation among independent variables, the $\Delta(R^2)$ for each variable would not depend on the order of entry and, of course, dropping or entering variables would not change the coefficients of the variables in the regression (Section 3.1).

The values of F_{in} and F_{out} were then changed to 1.2 and the procedures repeated (Tables 4.17 and 4.18).

The stepwise procedure yielded

$$\hat{Y}_R = 283 + \underset{(2.16)}{4.44}\,X_1 - \underset{(11.6)}{40.4}\,X_6$$

with only X_6 significant at the .05 level. Note that the coefficients of X_1 and X_6 have changed.

The results of the backward elimination are quite different. Here

$$\hat{Y}_R = -122 + \underset{(1.19)}{8.32}\,X_1 + \underset{(2.50)}{7.34}\,X_2 - \underset{(.00596)}{.00897}\,X_3 + \underset{(2.78)}{9.89}\,X_4$$

with the coefficients of X_1, X_2, and X_4 significant at the .05 level ◀

This example is especially interesting in view of a practice of some researchers. Faced with a multitude of possible variables in a data bank, they may apply the stepwise procedure routinely to sift through this data bank and may look only at the final equation. The temptation to interpret the resulting coefficients in terms of cost or economic or sociometric response then becomes irresistable, especially if high significance is present. We cannot help but wonder how many theories had their birth in this manner.

On the other hand, if prediction is the sole object and all correlations are expected to continue, the final choice of variables may not be critical. It must be remembered, however, that sample performance was the criterion for selection. Variables may have been deleted merely because they did not have sufficient range in the sample to produce "significant" coefficients. The selected equation may not predict accurately.

Research in which little a priori knowledge is used beyond a list of candidate independent variables has been called exploratory. Unkind critics may prefer the term "fishing expedition"; although fishing expeditions often do catch fish, in this case the fish should be carefully examined before ingesting.

If the number of variables to choose from is large, exploratory research with selection procedures may lead to entirely spurious results. Tests of significance on individual variables (t tests or partial F tests) become misleading because they assume a test on a prespecified hypothesis. If a selection procedure reduces 30 potential variables to five, it is selecting one hypothesis from among[1] 142506 such hypotheses. It becomes relatively easy to find a hypothesis with significant sample coefficients.

Armstrong [5] provides an amusing hypothetical example of a researcher who has 31 data points and 30 variables. The researcher then selects an equation by using the stepwise procedure and finds an equation with eight variables and eight significant coefficients; his $\bar{R}^2$ is 0.85. Subsequently he learns that a mistake in labeling data had occurred and that the data he had analyzed had actually been generated by random numbers. Consequently some researchers feel that the number of data points should be about 10 or more times the number of variables before selection begins. The examples in this book violate such principles, but the author hides behind academic license.

It is desirable to make maximum possible use of a priori analysis to avoid this pitfall of selection procedures. In particular, a priori analysis would include a preselection of a small number of variables relative to the number of data points. This selection should be based on theoretical grounds and should include specification of the direction of relationships (signs or parameters) for the purposes of hypothesis testing.

If exploratory research is unavoidable, a split sample technique is suggested [5]; this means that hypothesis testing is done on fresh data and not on data that were used to develop the hypothesis. The selection procedure is now viewed not as a series of tests but simply as a means of ordering and analyzing the data. Sample data are split in two parts: selection is done on one part and testing on the other. If there are not enough data for splitting, a technique called Monte Carlo experiments [6] for obtaining approximate distributions for estimators is suggested.

[1]There are $\binom{30}{5}$ or $30!/5!25! = 142506$ ways to choose five variables from 30.

EXERCISES

4.1 Refer to Table 2.2. Find the partial F values and partial correlation coefficients for variables X_1 and X_2 by using only the information in the table and the fact that there were 20 observations in the data.

4.2 Refer to Table 4.7. Using only the information in the table, find the sum of squares about regression that would result if variable X_3 were dropped. Check the result with Table 4.8. Repeat for X_1.

4.3 Refer to Tables 4.7 and 4.10. Employ the partial F test in (4.3) to test the null hypothesis $H_0 : \beta_3 = \beta_5 = \beta_6 = 0$ against the alternative that H_0 is not true. Use the .05 level of significance. Comment on the relation of this test to the procedure that actually eliminated the variables X_3, X_5, and X_6.

4.4 Review the problem description in Example 2.1. Try to imagine it as a real problem.

(a) Comment on the general usefulness of a selection procedure in this problem.
(b) Which selection procedure is preferable in this problem,—stepwise regression or backward elimination?
(c) Discuss the choice of F_{in} and F_{out}. What would your preference be?
(d) What reservations would you have regarding the usefulness of the model after selection?

4.5 Discuss the applicability of selection procedures in the problem in Exercise 2.1.

4.6 Three researchers had a friendly discussion of the best criteria for the addition or deletion of variables in the stepwise regression. One preferred to use partial correlation coefficient values in place of F_{in} and F_{out}. In this way, for example, by setting $r_{in} = .5$, he added only those variables that could reduce the unexplained sum of squares by 25% or more. He thought that this was a better criterion than the partial F test because in the F test he would have had to pick compromise F values in view of changing degrees of freedom. The second researcher preferred to use t values for entry and deletion; for example, if the value t_{in} were set at 2.3, it would mean that if a regression with the candidate variable i were run its t value (b_i/S_{b_i}) would have had to exceed 2.3 for entry. He claimed that this procedure dealt directly with the significance of coefficients and therefore preferred it to the other two. The third researcher claimed that all three methods were equivalent. For any F_{in} and F_{out} values r_{in}, r_{out}, t_{in}, and t_{out} values could be found and the selections would proceed identically for each of the three methods. Evaluate the merits of each researcher's argument.

4.7 A professor of industrial psychology believed that she had isolated 13 factors that determined job satisfaction. For a randomly selected sample of 15 workers, she obtained scores on the factors by questionnaire. The scores ranged from zero to nine. She also obtained a measure of job satisfaction. Job satisfaction was depicted on a scale with 100 as standard, 70 as extreme dissatisfaction, and 130 as extreme satisfaction. The data, as punched on cards by a graduate assistant, appears in Table 4.19.

(a) The professor decided to use backward elimination to reduce the number of factors. She used $F_{out} = 3$ on the grounds that a high F value would guarantee

Table 4.19 Factors and Satisfaction Scores for Exercise 4.7

Observation	Factors													Satisfaction
	1	2	3	4	5	6	7	8	9	10	11	12	13	score
1	5	7	4	3	0	1	2	5	0	5	7	8	6	93
2	7	5	6	7	8	9	3	2	4	0	5	6	1	71
3	3	5	2	3	4	2	5	6	3	2	4	5	7	100
4	9	4	5	6	3	1	4	5	7	8	6	3	0	87
5	3	2	5	4	2	8	9	7	3	2	5	3	2	124
6	2	4	5	3	6	8	7	5	2	3	5	0	9	90
7	8	5	3	1	5	6	8	0	2	3	5	6	8	88
8	1	7	6	3	2	1	4	5	0	3	1	3	2	96
9	3	5	6	9	8	2	5	0	6	7	8	6	9	112
10	2	5	3	6	4	8	9	5	0	2	3	5	6	97
11	2	3	5	6	8	0	2	3	4	5	6	8	7	94
12	7	5	6	2	3	1	4	5	7	8	6	9	0	101
13	3	5	6	8	4	7	9	8	6	5	3	2	5	102
14	2	4	4	8	9	7	2	0	3	5	6	8	5	122
15	4	4	5	8	9	2	3	5	4	6	7	8	5	90

the exclusion of factors with doubtful causality on job satisfaction. Run the backward elimination on a computer program. If your computer program has no backward elimination option, perform separate runs to affect the same result. Comment on the professor's reasoning and on the results.

(b) The professor presented her results at a conference on industrial job satisfaction. One member of her audience subsequently ran her data on a stepwise procedure using $F_{in} = F_{out} = 3$. He was surprised at the results. Perform the stepwise procedure. Discuss the results and compare them with the results of (a).

(c) Two years later a student approached the professor and told her the following story. The graduate student (since left for parts unknown) who was supposed to punch the questionnaire values on computer cards had lost the questionnaires. He substituted for factor scores whatever numbers came into his head. For the satisfaction number he used the formula $100 + z \cdot 10$, where z was a random normal deviation (he used column 1, page T-87 of *Experimental Statistics*[1] for values of z). Comment.

4.8 Construct a flow diagram (along the lines of Figure 4.1) for a "backward stepwise" procedure that allows variables to reenter after they have been eliminated. Comment on its usefulness.

4.9 Consider the data in Table 4.20

(a) Find all possible regressions with 1, 2, 3, and 4 independent variables. Plot maximum R^2 and $\bar{R}^2$ against the number of variables.

[1]*Experimental Statistics, Handbook 91*, United States Department of Commerce, National Bureau of Standards.

Table 4.20 Data for Exercise 4.9

X_1	X_2	X_3	X_4	Y
34	5	40	21	37
28	8	20	25	16
29	8	20	24	21
21	9	15	24	25
32	3	38	27	10
36	3	37	20	40
25	6	17	24	13
32	3	25	24	16
23	8	18	24	10
28	7	22	19	40
25	7	19	20	26
30	3	25	19	34

(b) Run forward selection and backward elimination procedures on the data ($F_{in} = .001$, $F_{out} = .001$). Compare the results with (a) above.

(c) Run the stepwise selection procedure with $F_{in} = F_{out} = 1$. Compare the results with (a) above.

4.10 Comment on the following quote:

A case in point . . . is that of prices of 15 . . . makes of cars regressed . . . on 6 . . . attributes giving respective values of t . . . none significant . . . at the .05 probability level . . . . Our procedure was to regress on the 2 variables with highest t's in the original set, namely variables numbered 2 and 6 (engine power and interior comfort) . . . t_2 is now overwhelmingly significant. Variable 6 . . . is (now) *formally* significant at the .02 level. While such probabilistic inference for variable 6 is no longer valid because of *ex post* selection, it seems reasonable to infer significance and to regard the variable as a causative factor. (From R. C. Geary and C. E. V. Leser, "Significance Tests in Multiple Regression," *The American Statistician*, February 1968.)

REFERENCES

1. N. R. Draper and H. Smith, *Applied Regression Analysis*. New York: Wiley, 1966, Chapter 6.

2. Normal Nie, Dale H. Bent, and C. Hadlai Hull, *Statistical Package for the Social Sciences*. New York: McGraw-Hill, 1970.

3. Yoel Haitovsky, "A Note on the Maximization of $\bar{R}^2$," *The American Statistician*, February 1969.

4. Moshe Weiss, "Letter to the Editor," *The American Statistician*, June 1970.

5. J. Scott Armstrong, "How to Avoid Exploratory Research," *Journal of Advertising Research*, 27–30, August 1970.

6. A. Ando and G. M. Kaufman, "Evaluation of an Ad Hoc Procedure for Estimating Parameters of Some Linear Models," *Review of Economics and Statistics*, **48**, 334–340, 1966.

chapter 5

Transformation and dummy variables

5.1 INTRODUCTION

Some functions that are actually nonlinear may be fitted by using linear regression, provided that they can be transformed into linear functions.[1] Examples are[2]

$$Y_R = \beta_0 + \beta_1 X_1 + \beta_{2.2} X_2^2 + \beta_{1,2} X_1 X_2$$

$$Y_R = \beta_0 X_1^{\beta_1} X_2^{\beta_2} X_3^{\beta_3}$$

$$Y_R = \left(1 - \beta_0 e^{\beta_1 X_1}\right)^{-1}$$

In this chapter we deal first with some common transformation methods. The use of categorical or "indicator" variables is discussed subsequently.

5.2 POLYNOMIAL REGRESSION

The order of a polynomial relationship is the highest degree present; for example, $X_1^2 X_2^3$ is a fifth-degree term and X_2^2 is a second-degree term. Again, for example, the relationship

$$Y_R = \beta_0 + \beta_1 X_1 + \beta_2 X_2 + \beta_{1.2} X_1^2 + \beta_{2.2} X_2^2 + \beta_{1,2} X_1 X_2$$

is a second-order relationship. Subscript notation must be somewhat modified for use in polynomials. The subscript $\beta_{j.p}$ means that this is the parameter of X_j raised to the pth power. The subscript $\beta_{j,i}$ refers to the

[1]Some humorist has called such functions "nonintrinsically nonlinear."

[2]The constant e is equal to 2.7182818 . . . Notation for the coefficients of the polynomial is explained in Section 5.2.

presence of variables X_j and X_i (which are multiplied). A term such as X_1X_2 can be used in an attempt to represent "interaction" between X_1 and X_2.

A least squares fit on a polynomial by means of multiple linear regression presents no new difficulties. The first independent variable in the polynomial above is still X_1, the second one is still X_2, the third is X_1^2, the fourth is X_2^2, and the fifth, X_1X_2.

In other words, we could define new variables T_1, T_2, T_3, T_4 and T_5, where $T_1 = X_1$, $T_2 = X_2$, $T_3 = X_1^2$, $T_4 = X_2^2$, and $T_5 = X_1X_2$. The polynomial is now a linear relationship:

$$Y_R = \beta_0 + \beta_1 T_1 + \beta_2 T_2 + \beta_{1.2} T_3 + \beta_{2.2} T_4 + \beta_{1,2} T_5$$

▶ **Example 5.1** A researcher had 13 data points for a dependent variable Y and for a single independent variable X_1 (Table 5.1).

Table 5.1 Thirteen Data Points

Y	10	16	28	30	40	50	60	70	70	61	70	79	80
X_1	1	1	3	3	4	5	6	7	8	9	10	11	16

He plotted the data (Figure 5.1) and found support for his theory that the response of Y to X_1 should exhibit diminishing returns[1] (slope). He decided to fit the polynomial model:

$$Y = \beta_0 + \beta_1 X_1 + \beta_{1.2} X_1^2 + \beta_{1.3} X_1^3 + \epsilon$$

The SPSS regression program [1] was used to perform the regression. Three independent variables, $T_1 = X_1$, $T_2 = X_1^2$, and $T_3 = X_1^3$, were supplied by a transformation feature of the program. Table 5.2 gives the results (see Example 4.3, Table 4.4, for an explanation of some features of the SPSS printout).

The regression found was $\hat{Y}_R = -1.76 + 13.7X_1 - .748X_1^2 + .0132X_1^3$, with a coefficient of determination of .962. The partial F of b_3, however, seemed low, and, in addition, the ratio of the number of data points to the number of parameters being fitted was low.

The analyst wished to test the null hypothesis $H_0 : \beta_3 = 0$ against $H_1 : \beta_3 \neq 0$ at the .01 level of significance. Because $F_{.01}$ is 10.56 at $\nu_1 = 1$

[1]This function is called concave.

Table 5.2 Regression with the New Independent Variables T_1, T_2, and T_3

MULTIPLE R	.98072		F	SIGNIFICANCE
R SQUARE	.96180		75.54028	.000
STD DEVIATION	5.40637			
			SUM OF	MEAN
ANALYSIS OF VARIANCE		DF	SQUARES	SQUARE
REGRESSION		3	6623.86355	2207.95452
RESIDUAL		9	263.05953	29.22884

VARIABLES IN THE EQUATION

VAR	B	STD ERROR B	F / SIGNIFICANCE	BETA / ELASTICITY
T_1	13.685662	3.3610131	16.580255	2.4749536
			.003	1.73132
T_3	.13179617E-01	.19155281E-01	.47340034	.6152235
			.509	.16816
T_2	−.74792296	.48468068	2.3812354	− 2.2113641
			.157	−.86507
(CONST)	− 1.7575775	6.0790103	.83591708E-01	
			.779	

and $\nu_2 = n - m = 13 - 4 = 9$, the null hypothesis cannot be rejected,[1] for the partial F of b_3 is .473.

When the variable X_3 was dropped, the result was $\hat{Y}_R = 1.33 + 11.5X_1 - .418X^2$, with an R^2 of .960 as shown in Table 5.3.

Because of multicollinearity, the coefficients of the relationship changed even though R^2 decreased only slightly. Since the SIGNIFICANCE of the partial F value for T_2 or X_1^2 is now .000, it is obvious that the null hypothesis $H_0 : \beta_2 = 0$ (or Y_R is really a straight line) would be rejected. Figure 5.1 shows the final regression. Note that to some extent the analyst has permitted the data to select the model for him. He should be familiar with the cautions discussed at the end of Chapter 4. ◀

Polynomials have some valuable features as a tool for regression analysis; unfortunately, there is a list of drawbacks. First let us consider the good points. The transformations used to convert a polynomial to a linear relationship were of a trivial sort: we merely substituted new independent

[1]The SIGNIFICANCE value of the partial F for b_3 is .509 (from Table 5.2). This means that $\Pr(F \geqslant .473) = .509$ if H_0 is true. In turn, this tells us that .473 does not fall in the critical region at the .01 level of significance, hence no additional tables needed to be used.

Table 5.3 Regression Equation After the Removal of T_3 ($T_3 = X_1^3$)

MULTIPLE R	.97969	F	SIGNIFICANCE
R SQUARE	.95979	119.35917	.000
STD DEVIATION	5.26210		

ANALYSIS OF VARIANCE	DF	SUM OF SQUARES	MEAN SQUARE
REGRESSION	2.	6610.02661	3305.01331
RESIDUAL	10.	276.89647	27.68965

VARIABLES IN THE EQUATION

VAR	B	STD ERROR B	F / SIGNIFICANCE	BETA / ELASTICITY
T_1	11.522885	1.1580159	99.013268	2.0838309
			.000	1.45771
T_2	−.41822302	.70829208E-01	34.865115	− 1.2365490
			.000	−.48373
(CONST)	1.3286880	3.9934232	.11070198	
			.746	

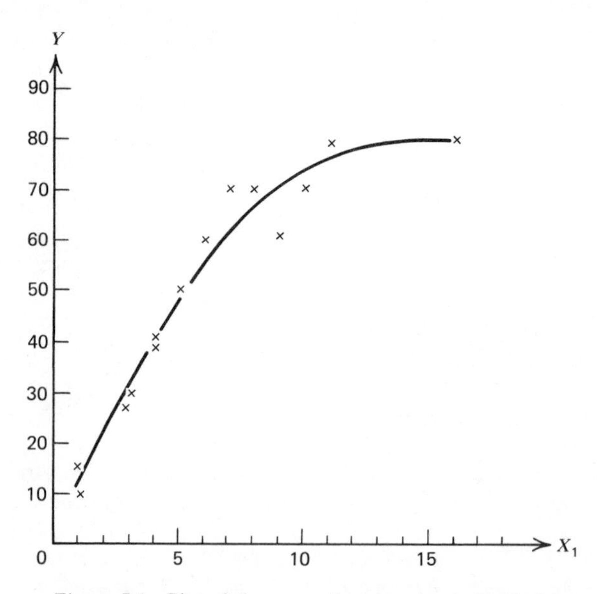

Figure 5.1 Plot of the regression equation in Table 5.3.

variables into a linear relationship.[1] Therefore the formulas for least squares estimation still apply and the parameter estimators are still unbiased minimum variance estimators. Polynomials are a flexible tool in the sense that they can give many different "shapes" to a relationship Y_R and allow the use of interaction terms. In practice, third-order polynomials are generally the highest order used. The coefficients of higher order terms are difficult to interpret. Polynomials are often used to represent a relationship when nonlinearity or curvature is suspected. The presence of curvature is then examined by testing the coefficients of higher order terms as in the preceding example.[2]

As usual multicollinearity causes problems. The correlation between two variables is, as will be recalled from Chapter 1, a measure of the degree of a linear relationship between those variables. In Figure 5.2 X is plotted against X^2 and X^2 against X^3. If X is in the range [10, 14], then X^2 is in the range [100, 196] and X^3 is in the range [1000, 2744]. The correlation coefficients among the three variables X, X^2, and X^3 within that particular range for X would obviously be high because their interrelationships are nearly linear. Therefore the standard errors of regression coefficients are large if the variation in X is small. Thus it is possible that two or more different polynomials give nearly the same fit; for example, $Y_R = \beta_0 + \beta_{1.2}X_1^2 + \beta_{1.3}X_1^3$ might give about the same coefficient of determination as $Y_R = \beta_0 + \beta_1 X_1 + \beta_{1.4}X_1^4$.

Some researchers may argue that they are interested only in prediction and not in the coefficients themselves. Even here they are on dangerous ground. A polynomial may be quite unpredictable outside the range for which it was fitted.

Figure 5.3*a* shows a curve a researcher expects from theory, the data points he gathered are in Figure 5.3*b*, and the polynomial he fitted is in Figure 5.3*c*. Also, because the researcher has already fitted a five-parameter curve to only 10 data points, we could urge him one step further: use a 10-parameter polynomial and obtain a perfect R^2 of one[3] (all the X_1's in the data are different).

With respect to the danger of extrapolating polynomials, we could argue that one is safe as long as only interpolation is attempted. As shown in

[1]Nonlinear curves such as $Y_R = \beta_0 + \beta_1(1/X_1) + \beta_2 \log(X_2)$ can be handled by the same principles. The first independent variable is $1/X_1$ and the second is $\log(X_2)$.

[2]We should be careful here; for example, it is possible to have an S shaped curve where, when $Y_R = \beta_0 + \beta_1 X_1 + \beta_{1.2}X_1^2$ is fitted, the coefficient of X_1^2 is not significant, yet when $Y_R = \beta_0 + \beta_1 X_1 + \beta_{1.2}X_1^2 + \beta_{1.3}X_1^3$ is fitted the coefficients of X_1^2 and X_1^3 are both significant. The reader should attempt to explain this example.

[3]If the advice does not strike the reader as humorous, either his sense of humor is deficient or he should reread Chapter 3 or both.

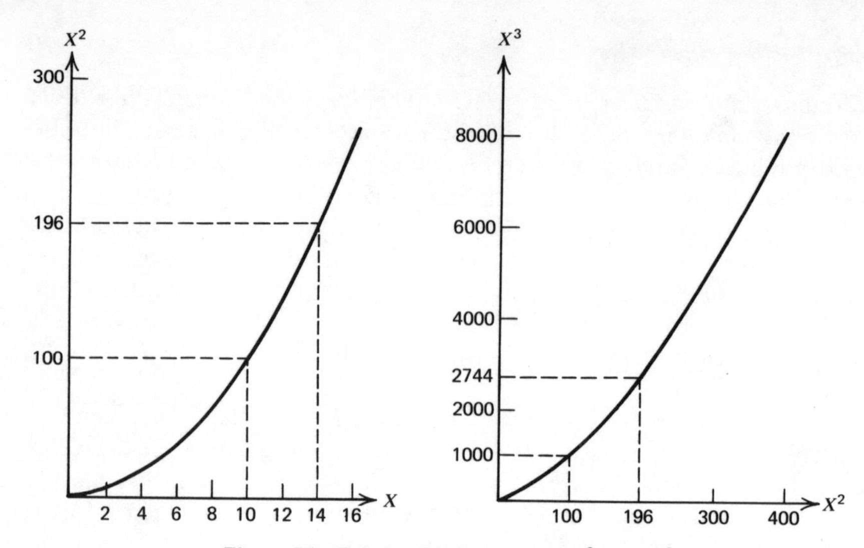

Figure 5.2 Relationship between X, X^2, and X^3.

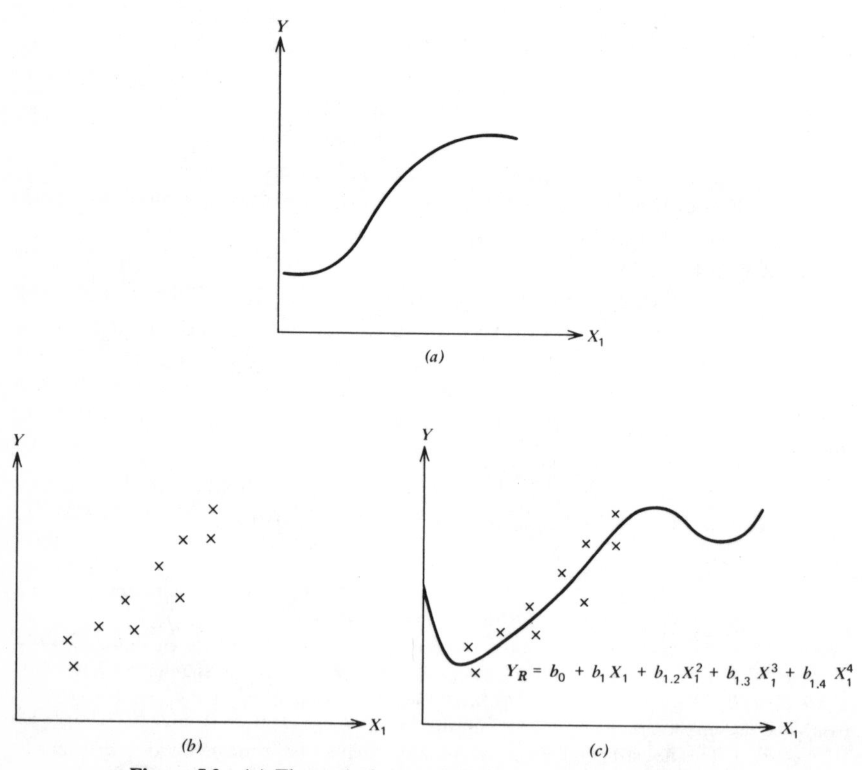

Figure 5.3 (*a*) Theoretical curve; (*b*) data points; (*c*) curve fitting.

Section 3.5, however, if there is more than one independent variable present we may choose all independent variables within the range of those appearing in the data and still be extrapolating.

5.3 LOGARITHMIC TRANSFORMATIONS

By taking the logarithms of both sides of a nonlinear relationship we can often transform it so that linear regression can be used. Fortunately a great many theoretically important nonlinear relationships can be transformed; unfortunately the nature of the error term that was originally assumed in Chapter 2 must be altered rather drastically; for example, consider the following:

$$Y_R = \beta_0 X_1^{\beta_1} X_2^{\beta_2} \cdots X_q^{\beta_q} \tag{5.1}$$

which is known as a *constant elasticity function*[1] and is widely used for demand analysis both in economics and in marketing.

Assume for the moment that the error term has the somewhat peculiar form shown in (5.2):

$$Y = \beta_0 X_1^{\beta_1} X_2^{\beta_2} \cdots X_q^{\beta_q} e^{\epsilon} \tag{5.2}$$

Taking natural[2] logarithms of both sides of the equation, we have

$$\ln Y = \ln \beta_0 + \beta_1 \ln X_1 + \beta_2 \ln X_2 + \cdots + \beta_q \ln X_q + \epsilon \tag{5.3}$$

If we let $\ln(Y)$ be the dependent variable and $\ln X_1, \ln X_2, \ldots, \ln X_q$ be the independent variables, standard multiple linear regression may be used to find the estimators of $\ln \beta_0, \beta_1, \ldots, \beta_q$.

It may now be asked why the error term was assumed to have the exponential form of (5.2). The answer is simple: we could not have made a logarithmic transformation[3] with an additive error term of the type present in (2.4). The opportunistic argument is sometimes made that we probably did not have much reason to assume an additive term in the first place.[4]

We should not really have it both ways, although some researchers try first a linear and then a constant elasticity relationship; they then compare

[1]See Appendix A, Note 5.1, for a derivation of the elasticity.

[2]For a brief review of natural logarithm operations see Appendix A, Note 5.2. Natural logarithms are often handier; in this application, however, common logarithms could have been used just as well.

[3]It is possible to make parameter estimates with an additive error term, but we must use nonlinear optimization techniques [2].

[4]Methods of checking error term assumptions are discussed in Chapter 6.

the R^2's obtained. This is somewhat like comparing the traditional apples and oranges because the first R^2 tells us what proportion of the variation in Y was explained by the relationship and the second tells us what proportion of the variation in $\ln Y$ was explained.

The use of significance tests on coefficients obtained by searching for the best form of a relationship is also a sin. As explained in Section 4.5, we cannot take at face value the significance of coefficients after using a selection procedure. Similarly, searching for a relationship with the best "significances" amounts to searching actively for a hypothesis that can be rejected with the data. It is easy to obtain spurious significance, especially if the number of independent variables is not small compared with the number of data points.

It should also be remembered that no zero or negative values are allowed if (5.3) is used for obtaining regression coefficients. The logarithm of zero is $-\infty$ and logarithms of negative numbers are not defined. The presence of zero or negative values in the data could argue against the use of the functional form (5.2).

Something should be said about the statistical properties of the estimators in (5.3). If ϵ is normally distributed with mean zero and a constant standard deviation, the estimators $b_0, b_1, \ldots, b_q$ will be unbiased minimum variance estimators for the parameters $\ln \beta_0, \beta_1, \ldots, \beta_q$; because (5.3) is a linear regression, we retain the desirable properties of a standard linear regression. Note that the above would not be true if the error term were, in reality, additive. A further point to remember is that taking the antilog of b_0 does *not* give an unbiased estimator for β_0.

Other logarithmic transformations are listed in Table 5.4.[1] It is, of course, only a small sample of the types of transformations possible.

Table 5.4 Three Transformations

Model	Transformed Relationship
1. $Y = \exp(\beta_0 + \beta_1 X_1 + \cdots + \beta_q X_q)e^{\epsilon}$	$\ln Y = \beta_0 + \beta_1 X_1 + \cdots + \beta_0 X_q + \epsilon$
2. $Y = \beta_0 \beta_1^{X_1} \cdots \beta_q^{X_q} e^{\epsilon}$	$\ln Y = \ln \beta_0 + \ln \beta_1(X_1) + \cdots + \ln \beta_q(X_q) + \epsilon$
3. $Y = \dfrac{1}{1 + \beta_0 \exp(\beta_1 X_1 + \cdots + \beta_q X_q + \epsilon)}$	$\ln(\frac{1}{Y} - 1) = \ln \beta_0 + \beta_1 X_1 + \cdots + \beta_q X_q + \epsilon$

In general, whenever the parameters are not transformed (as in Model 1, Table 5.4) and whenever we can assume that the error term ϵ is normally

[1]See Appendix A, Note 5.2, for transformation methods for Models 1 and 3.

distributed with mean zero and constant variance, the regression coefficients obtained will be minimum variance unbiased estimators of the parameters. When parameters must be transformed (as in Model 2, Table 5.4), the regression coefficients obtained will be unbiased estimators of the *transformed* parameters (i.e., unbiased estimates of $\ln \beta_0$, $\ln \beta_1, \ldots, \ln \beta_q$ will be found for Model 2).

▶ **Example 5.2** The following nonlinear curves were fitted to the data in Example 5.1:

$$\text{(i)} \quad Y = e^{\beta_0 + \beta_1 x_1} e^{\epsilon}$$

$$\text{(ii)} \quad Y = \beta_0 X_1^{\beta_1} e^{\epsilon}$$

Actually, preliminary analysis[1] would have shown that (i) was probably not a suitable candidate. When possible, nonlinear curves should be examined to see if they fulfill a priori expectations. In exploratory analysis curves should at least conform to "shapes" suggested by the data.

Equation (i) was transformed as follows:

$$Y = e^{\beta_0 + \beta_1 X_1} e^{\epsilon}$$

$$\ln Y = \beta_0 + \beta_1 X_1 + \epsilon$$

The fitted relationship[2] was

$$\widehat{(\ln Y)}_R = \underset{(.196)}{2.97} + \underset{(.0255)}{.125} X_1$$

with an R^2 of .687 and an F value of 24.1 which is significant at the .01 level. Therefore

$$\hat{Y}_R = e^{2.97 + .125 X_1}$$

Given the form of the model, the estimators for the parameters are unbiased. Equation (ii), a special case of (5.2), was also transformed by taking logarithms of both sides:

$$Y = \beta_0 X_1^{\beta_1} e^{\epsilon}$$

$$\ln Y = \ln \beta_0 + \beta_1 \ln X_1 + \epsilon$$

The fitted relationship was

$$\widehat{(\ln Y)}_R = \underset{(.102)}{2.61} + \underset{(.0568)}{.736} \ln X_1$$

[1]See Appendix A, Note 5.3.

[2]Recall that the values in brackets under the coefficients are the standard errors of those coefficients.

with an R^2 of .939 and an F value of 168.0 which is certainly significant at the .01 level. Transforming back to the original form, we have

$$\hat{Y}_R = e^{2.61}X_1^{.736}$$
$$= 13.6X_1^{.736}$$

Note that the estimate of 13.6 for β_0 is biased but that the estimate 2.61 for $\ln \beta_0$ is not.

Table 5.5 gives the calculated values and residuals for curves (i) and (ii) provided by an SPSS program. The curves have distinct groups of negative and positive residuals, and when plotted the residuals exhibit a distinct

Table 5.5 Residuals for Example 5.3

Observation	Y value	Y estimate	Residual
	$Y = e^{\beta_0+\beta_1 X_1} e^{\epsilon}$		
1	2.302585	3.083843	−.7912578
2	2.772589	3.093843	−.3212542
3	3.332205	3.344313	−.1210881E-01
4	3.401197	3.344313	.5688406E-01
5	3.688879	3.469549	.2193309
6	3.912023	3.594784	.3172393
7	4.094345	3.720019	.3743257
8	4.248495	3.845254	.4032412
9	4.248495	3.970489	.2780060
10	4.110874	4.095724	.1514939E-01
11	4.248495	4.220960	.2573558E-01
12	4.369448	4.346195	.2325299E-01
13	4.382027	4.972371	−.5903442
	$Y = \beta_0 X_1^{\beta_1} e^{\epsilon}$		
1	2.302585	2.607228	−.3046428
2	2.772589	2.607228	.1653609
3	3.332205	3.415644	−.8343988E-01
4	3.401197	3.415644	−.1444701E-01
5	3.688879	3.627336	.6154351E-01
6	3.912023	3.791537	.1204863
7	4.094345	3.925698	.1686461
8	4.248495	4.039131	.2093646
9	4.248495	4.137390	.1111052
10	4.110874	4.224061	−.1131871
11	4.248495	4.301591	−.5309549E-01
12	4.369448	4.371725	−.2277104E-01
13	4.382027	4.647444	−.2654174

pattern. This is often an indication that a better fitting curve might be found.[1] ◀

▶ **Example 5.3** The following equation was fitted to the data in Example 2.1:

$$Y_R = \beta_0 X_1^{\beta_1} X_2^{\beta_2} X_3^{\beta_3} X_4^{\beta_4} X_5^{\beta_5} X_6^{\beta_6}$$

After transformation

$$(\ln Y)_R = \ln \beta_0 + \beta_1 \ln X_1 + \beta_2 \ln X_2 + \beta_3 \ln X_3 + \beta_4 \ln X_4 + \beta_5 \ln X_5 + \beta_6 \ln X_6$$

The SPSS printout is given in Table 5.6.

Table 5.6 Regression Printout for the Data in Example 2.1 (after a logarithmic transformation)

MULTIPLE *R*	.94448	*F*	SIGNIFICANCE
R SQUARE	.89204	17.90248	.000
STD DEVIATION	.04221		

ANALYSIS OF VARIANCE	DF	SUM OF SQUARES	MEAN SQUARE
REGRESSION	6.	.19136	.03189
RESIDUAL	13.	.02316	.00178

VARIABLES IN THE EQUATION

VAR	*B*	STD ERROR *B*	*F* / SIGNIFICANCE	BETA / ELASTICITY
ln(X_1)	.44079427	.16507950	7.1299355	.6801737
			.019	−.01220
ln(X_3)	.28337055	.71011990	.15923776	.0449636
			.696	.23005
ln(X_2)	.31572076	.11305619	7.7986164	.3225865
			.015	−.00328
ln(X_4)	−.84967300E-01	.44444174E-01	3.6548871	−.2536676
			.078	−.00510
ln(X_5)	.87743231E-01	.13759271	.40666465	.1130435
			.535	−.00175
ln(X_6)	−.89408565E-01	.17107627	.27313585	−.1192347
			.610	.04424
(CONST)	1.7920379	1.4805710	1.4649942	
			.248	

[1] See Appendix A, Note 5.3.

Because $\widehat{(\ln Y)}_R = 1.79 + .441 \ln X_1 + .316 \ln X_2 + .283 \ln X_3 - .0850 \ln X_4 + .0877 \ln X_5 - .0894 \ln X_6$,

$$\hat{Y}_R = e^{1.79} X_1^{.441} X_2^{.316} X_3^{.283} X_4^{-.0850} X_5^{.0877} X_6^{-.0894}$$

or

$$\hat{Y}_R = 6.00 X_1^{.441} X_2^{.316} X_3^{.283} X_4^{-.0850} X_5^{.0877} X_6^{-.0894}$$

The partial F values of the sample regression coefficients of $\ln X_3$, $\ln X_4$, $\ln X_5$ and $\ln X_6$ are not significant at the .05 level. In fact, when the backward elimination procedure (with $F_{out} = 1$) was applied, X_3, X_5, and X_6 were dropped as they were for the linear relationship in Example 4.4, Table 4.10.

It is interesting to compare the postselection linear relationship

$$\hat{Y}_R = 2.47 + 6.75X_1 + 3.09X_2 - .714X_4 \qquad R^2 = .899$$

with the postselection constant elasticity relationship

$$\hat{Y}_R = 11.6 X_1^{.556} X_2^{.298} X_4^{-.628}, \qquad R^2 = .887 \text{ (on transformed relationship)}$$

Table 5.7 gives predicted values of Y (or $\hat{Y}_R$) for each relationship.

Table 5.7 Comparison of Linear and Constant Elasticity Relationships

Observation	Actual Y value	Y estimated from the linear relationship	Y Estimated from the constant elasticity relationship
1	9.330	9.416	9.413
2	11.56	12.07	12.07
3	10.32	10.44	10.42
4	11.35	11.75	11.57
5	11.21	11.06	11.07
6	10.90	10.13	10.18
7	11.82	11.90	11.95
8	10.06	10.26	10.24
9	9.910	9.484	9.618
10	11.98	11.30	11.25
11	13.69	13.55	13.50
12	11.44	11.34	11.35
13	10.58	10.71	10.58
14	14.10	13.48	13.27
15	10.76	11.13	11.01
16	10.59	11.24	11.41
17	10.12	10.36	10.34
18	9.360	9.237	9.268
19	10.72	10.72	10.76
20	10.92	11.16	11.31

There is evidently not much advantage in one relationship over the other as far as fit is concerned. However, extrapolation of the linear relationship beyond the range of the data fitted could give different results from extrapolating the nonlinear relationship to the same point. Also, this example should caution against assuming that a good fit of a linear relationship proves linearity; we could simply be using data on a "flattish" portion of a curved surface. ◀

5.4 DUMMY (INDICATOR) VARIABLES

So far we have considered only virtually continuous variables such as dollar sales, dollar GNP, and temperature. In data analysis, however, *categorical* or *qualitative* variables appear (e.g., sex, race, disability) which may have no natural numerical scales. In addition, some variables such as age may have numerical scales, but we may wish to use only three categories, say, young, middle-aged, and old. Regression variables called *dummy*, *indicator*, or *binary* variables provide a flexible tool for handling categories in data. They extend considerably the power of regression to solve practical problems.

To introduce dummy variable methods we use the following hypothetical example. The annual expenditure per capita on a product is thought to vary linearly with disposable personal income within a certain narrowly defined consumer group. The level of expenditure, however, is thought to be different for each of three age categories: young (18–40 years), middle-aged (41–75 years), and old (76 years and over).

The model can now be represented as

$$Y = \beta_0 + \beta_1 X_1 + \delta_1 D_1 + \delta_2 D_2 + \epsilon \tag{5.4}$$

where Y = the annual expenditure on the product by a consumer,

X_1 = the consumer's annual disposable income,

$$D_1 = \begin{cases} 1 \text{ if the consumer is in the young category,} \\ 0 \text{ otherwise,} \end{cases}$$

$$D_2 = \begin{cases} 1 \text{ if the consumer is in the middle aged category,} \\ 0 \text{ otherwise.} \end{cases}$$

The dummy variables[1] D_1 and D_2 identify the consumer's age group; for

[1]In this chapter special notation (D_j's) is used for dummy variables in an attempt at clarity of presentation. Dummy variables are simply independent variables and the regular (X_j's) notation is used in subsequent chapters.

example, a consumer who is young would have $D_1 = 1, D_2 = 0$, one who is middle-aged would have $D_1 = 0, D_2 = 1$, and one who is old would have $D_1 = 0, D_2 = 0$. The δ notation for coefficients of dummy variables is used instead of β for convenience.

The question is asked, why not use D_3 to represent the elderly category? The answer is that it is both unnecessary and impossible to use as many dummy variables as there are categories.

The purpose of the dummy variables in this example is to allow a different level of consumption for each age category while keeping the same marginal response to annual disposable income (β_1). Table 5.8 presents a categorical breakdown of (5.4) to show that the equation fulfills this purpose.

Table 5.8 The Role of Dummy Variables in Changing an Equation

Category	Dummy variable values	Equation
Young	$D_1 = 1, D_2 = 0$	$Y = \beta_0 + \delta_1 + \beta_1 X_1 + \epsilon$
Middle-aged	$D_1 = 0, D_2 = 1$	$Y = \beta_0 + \delta_2 + \beta_1 X_1 + \epsilon$
Old	$D_1 = 0, D_2 = 0$	$Y = \beta_0 + \beta_1 X_1 + \epsilon$

Figure 5.4 illustrates the possibility that there is a different intercept for each age category. In effect, three different lines with the same slope are being fitted. This is not the same as separating the data points into the three categories and running separate regressions because the latter method would produce three different slopes.

Had D_3 been used as well, no solution could have been found: because every consumer has to fall into some category, $D_1 + D_2 + D_3$ is equal to

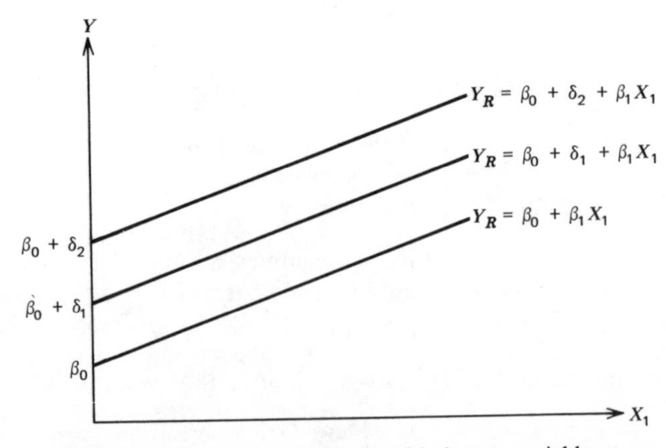

Figure 5.4 Changing the intercept with dummy variables.

one, which means that an exact linear relationship exists among the independent variables. This means that no unique estimators for the regression parameters can be found.[1] In general, the following rule holds:

Rule. A categorical variable with p categories will be represented by $p - 1$ dummy variables. (5.5)

Dummy variables can also be used to represent interaction effects; for example, suppose it were hypothesized that the marginal response to income is lower in the young age category than in the other two.

Equation 5.4 would now become

$$Y = \beta_0 + \beta_1 X_1 + \delta_1 D_1 + \delta_2 D_2 + \delta_3 D_1 X_1 + \epsilon \tag{5.6}$$

The new variable $D_1 X_1$ could change the slope of the line for the young category as in Figure 5.5.

The following examples illustrate some of the possible applications of dummy variables.

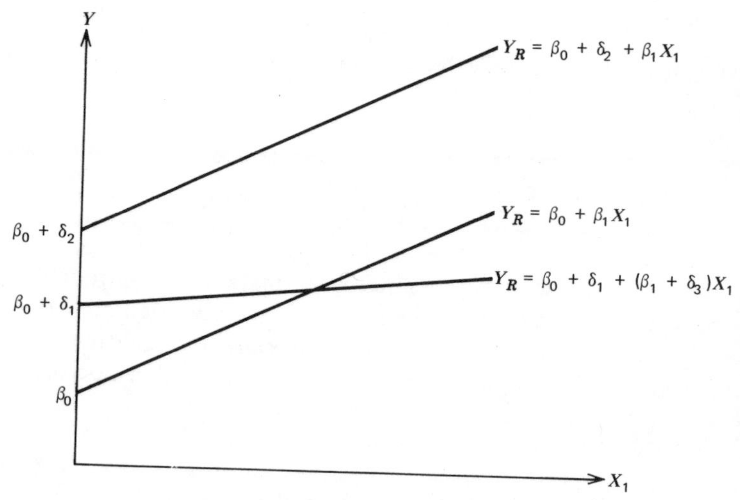

Figure 5.5 Changing the slope with a dummy variable.

▶ **Example 5.4** An economist wished to study the growth of the demand for calgacite, an industrial chemical. He had demand figures for the last 15 years. Table 5.9 reproduces his data and Figure 5.6 shows the plot of the observations.

[1] This is the extreme case of multicollinearity.

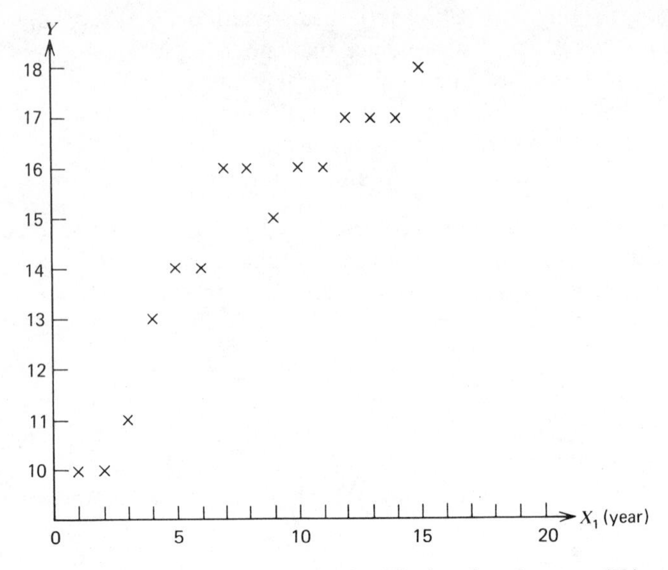

Figure 5.6 Demand for calgacite (Y) plotted against year (X_1).

Table 5.9 Demand for Calgacite Recorded for 15 Years

Y = demand in thousands of tons	10	10	11	13	14	14	16	16	15	16	16	17	17	17	18
X_1 = Year	1	2	3	4	5	6	7	8	9	10	11	12	13	14	15

The economist knew that the ninth year marked the introduction of major changes in a competing chemical. He hypothesized that the new competing product increased or decreased the market for calgacite but that the rate of growth for the market remained unchanged (because the needs of users grew with time at the same rate).

His model was

$$Y_R = \beta_0 + \beta_1 X_1 + \delta_1 D_1$$

where

$$D_1 = \begin{cases} 1 & \text{if} \quad X_1 \geqslant 9 \\ 0 & \text{if} \quad X_1 < 9 \end{cases}$$

An SPSS printout for this regression is presented in Table 5.10. The fitted

Table 5.10 Regression Printout for Example 5.4

MULTIPLE R	.95939		F	SIGNIFICANCE
R SQUARE	.92043		69.40490	.000
STD DEVIATION	.79507			
			SUM OF	MEAN
ANALYSIS OF VARIANCE		DF	SQUARES	SQUARE
REGRESSION		2	87.74762	43.87381
RESIDUAL		12	7.58571	.63214

VARIABLES IN THE EQUATION

VAR	B	STD ERROR B	F / SIGNIFICANCE	BETA / ELASTICITY
X_1	.75714286	.95029534E-01	63.480226	1.2975809
			.000	.41299
D_1	− 2.1071429	.82297990	6.5555556	−.4169846
			.025	−.06705
(CONST)	9.5928571	.51174970	351.38301	
			.000	

linear relationship is

$$\hat{Y}_R = \underset{(.512)}{9.59} + \underset{(.0950)}{.757X_1} - \underset{(.823)}{2.11D_1}$$

or alternatively

$$\begin{aligned} \hat{Y}_R &= 9.59 + .757X_1 \quad \text{if } X_1 < 9 \\ &= 7.49 + .757X_1 \quad \text{if } X_1 \geqslant 9 \end{aligned}$$

This relationship is plotted in Figure 5.7.

Suppose that the economist wished to test the hypothesis that no change had occurred in the ninth year at the .05 level of significance. More specifically, $H_0 : \delta_1 = 0$ and $H_1 : \delta_1 \neq 0$. A two-tailed t test could be used but the partial F test is equivalent (review Section 4.2). The partial F value for the sample regression coefficient of D_1 is 6.56 (Table 5.10), whereas[1] $F_{.05}$ at $\nu_1 = 1$ and $\nu_2 = n - m = 15 - 3 = 12$ degrees of freedom is 4.75 (Table 3 in Appendix C). The economist would therefore reject the hypothesis that no change had occurred. ◀

[1]Table 3 is not necessary if the SPSS program is used. Table 5.10 tells us that $\Pr(F \geqslant 6.56) = .025$; it follows that 6.56 is in the critical region.

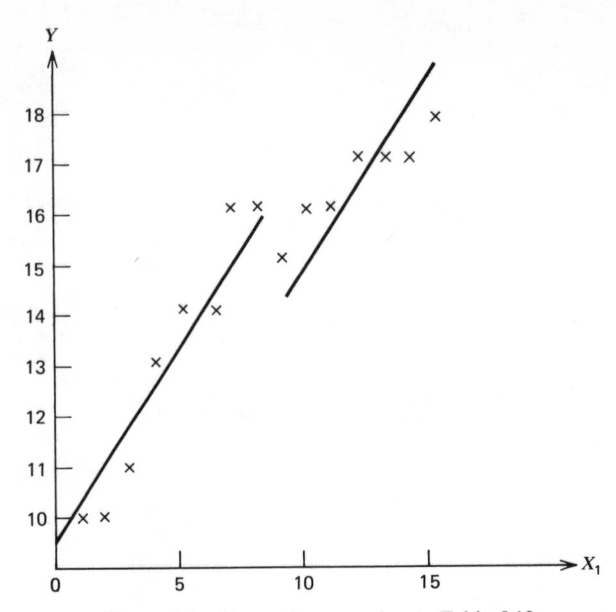

Figure 5.7 Plot of the regression in Table 5.10.

▶ **Example 5.5** Suppose that the economist believed that not only had the demand for calgacite changed as a result of the competing product but that the rate of growth also changed. His model would be

$$Y_R = \beta_0 + \beta_1 X_1 + \delta_1 D_1 + \delta_2 D_1 X_1$$

where

$$D_1 = \begin{cases} 1 & \text{if} \quad X_1 \geqslant 9 \\ 0 & \text{if} \quad X_1 < 9 \end{cases}$$

Table 5.11 gives the regression results. The fitted linear relationship would be:

$$\hat{Y}_R = 8.61 + .976 X_1 + 2.82 D_1 - .548 D_1 X_1$$

or

$$\begin{aligned} \hat{Y}_R &= 8.61 + .976 X_1 \quad \text{if } X_1 < 9 \\ &= 11.4 + .429 X_1 \quad \text{if } X_1 \geqslant 9 \end{aligned}$$

Table 5.11 Regression Printout for Example 5.5

MULTIPLE R	.98655		F	SIGNIFICANCE
R SQUARE	.97328		133.54206	.000
STD DEVIATION	.48125			

ANALYSIS OF VARIANCE	DF	SUM OF SQUARES	MEAN SQUARE
REGRESSION	3	92.78571	30.92857
RESIDUAL	11	2.54762	.23160

VARIABLES IN THE EQUATION

VAR	B	STD ERROR B	F SIGNIFICANCE	BETA ELASTICITY
X_1	.97619048	.74258514E-01	172.81308	1.6729817
			0.0000	.53247
D_1	2.8214286	1.1682448	5.8327103	.5583353
			.034	.08977
D_1X_1	−.54761905	.11741302	21.753271	− 1.3338619
			.001	−.20909
(CONST)	8.6071429	.37498712	526.84799	
			0.000	

This regression is plotted in Figure 5.8.

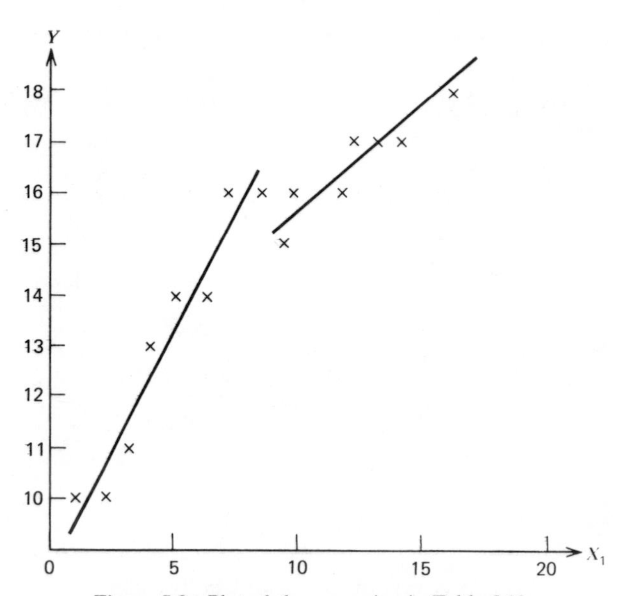

Figure 5.8 Plot of the regression in Table 5.11.

Note that this result could have been obtained by using two separate regressions: one for data points where $X_1 < 9$ and one for data points where $X_1 \geqslant 9$.

If the economist wished to test the hypothesis that the relationship is a single straight line, he would use $H_0 : \delta_1 = \delta_2 = 0$ against $H_1 : \delta_1, \delta_2$ are not both zero.

One method of doing this is the F test in (4.3). He could consider this problem a partial F test of dropping two variables (D_1 and D_1X_1) simultaneously. First the economist would run a regression with X_1 as the only independent variable (Table 5.12).

Table 5.12 Regression with X_1 the Only Independent Variable

MULTIPLE *R*	.93646		*F*	SIGNIFICANCE
R SQUARE	.87696		92.65716	0.000
STD DEVIATION	.94989			
			SUM OF	MEAN
ANALYSIS OF VARIANCE		DF	SQUARES	SQUARE
REGRESSION		1	83.60357	83.60357
RESIDUAL		13	11.72976	.90229

VARIABLES IN THE EQUATION

			F	BETA
VAR	*B*	STD ERROR *B*	SIGNIFICANCE	ELASTICITY
X_1	.54642857	.56766734E-01	92.657160	.9364617
			0.000	.29805
(CONST)	10.295238	.51613002	397.88224	
			0.000	

From (4.3)

$$F = \frac{\text{unexplained SS without } D_1 \text{ and } D_1X_1 - \text{unexplained SS with } D_1 \text{ and } D_1X_1}{\text{unexplained SS with } D_1 \text{ and } D_1X_1}$$

$$\times \frac{15-4}{2} = \frac{11.73 - 2.55}{2.55} \times \frac{11}{2} = 19.8$$

Because $F_{.05}$ (when $\nu_1 = 2, \nu_2 = 11$) is 3.98, the hypothesis of simple linearity, or no effect of competition, would be rejected at the .05 level of significance. ◀

▶ **Example 5.6** Suppose that the economist hypothesized that there was a change in growth in the middle of the ninth year but that there was no change in demand as a result of the competitive innovation. His model[1] would then be

$$Y_R = \beta_0 + \beta_1 X_1 + \delta_2 (X_1 - 8.5) D_1$$

where

$$D_1 = \begin{cases} 1 & \text{if} \quad X_1 \geqslant 9 \\ 0 & \text{if} \quad X_1 < 9 \end{cases}$$

Note that, as required, the line before competition, $Y_R = \beta_0 + \beta_1 X_1$, and the line after competition, $Y_R = \beta_0 + \beta_1 X_1 + \delta_2 (X_1 - 8.5)$, cross at $X_1 = 8.5$; for $\beta_0 + \beta_1 X_1$ to be equal to $\beta_0 + \beta_1 X_1 + \delta_2 (X_1 - 8.5)$, X_1 must be 8.5.

The regression printout is in Table 5.13.

Table 5.13 Regression Printout for Example 5.6

MULTIPLE *R*	.96996		*F*	SIGNIFICANCE
R SQUARE	.94082		95.38819	.000
STD DEVIATION	.68567			

ANALYSIS OF VARIANCE	DF	SUM OF SQUARES	MEAN SQUARE
REGRESSION	2	89.69165	44.84583
RESIDUAL	12	5.64168	.47014

VARIABLES IN THE EQUATION

VAR	*B*	STD ERROR *B*	*F* / SIGNIFICANCE	BETA / ELASTICITY
X_1	.81545988	.85254402E-01	91.489568	1.3975238
			.000	.44480
$(X_1 - 8.5)D_1$	−.59784736	.16613605	12.949495	−.5257749
			.004	.06658
(CONST)	9.1194716	.49553867	338.67586	
			.000	

[1]The model $Y_R = \beta_0 + \beta_1 X_1 + \delta_1 D_1 X_1$ would have been incorrect. See Appendix A, Note 5.4.

The fitted relationship is

$$\hat{Y}_R = 9.12 + .815X_1 - .598(X_1 - 8.5)D_1$$

or

$$\hat{Y}_R = 9.12 + .815X_1 \quad \text{if } X_1 < 9$$
$$= 14.2 + .218X_1 \quad \text{if } X_1 \geqslant 9$$

This relationship is plotted in Figure 5.9. ◀

The use of dummy variables is not restricted to incorporating categories into the regression; for example, a firm's assets may be described as a continuous variable in dollars. Some analysts, however, will break up the assets variable into categories; for example: less than \$500,000; \$500,000 up to \$1,000,000; \$1,000,000 up to \$1,500,000.

The advantage of using dummy variables to represent a continuous variable is that no assumption about the shape of the response need be made. For a continuous variable we must specify its effect as linear or transformed linear. Dummy variables will produce their own shape; for example, suppose that an analyst wished to describe earnings per share in terms of the asset categories above. Hence

$$Y_R = \beta_0 + \delta_1 D_1 + \delta_2 D_2$$

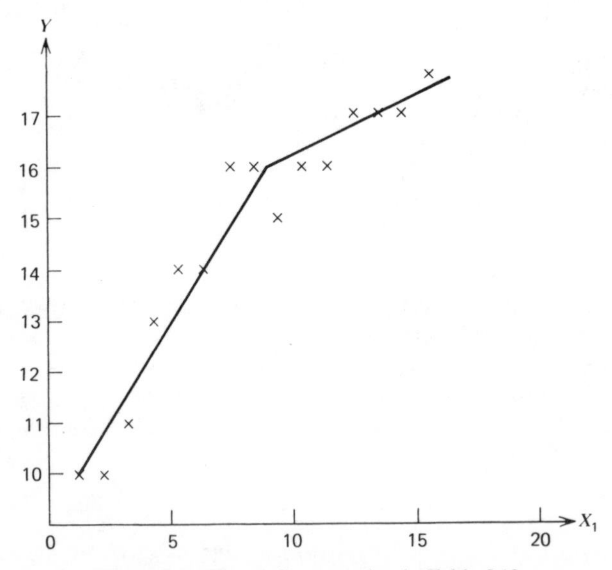

Figure 5.9 Plot of the regression in Table 5.13.

where

$$Y_R = \text{earnings per share},$$

$$D_1 = \begin{cases} 1 \text{ if } 0 \leqslant \text{assets} < 500{,}000, \\ 0 \text{ otherwise}, \end{cases}$$

$$D_2 = \begin{cases} 1 \text{ if } 500{,}000 \leqslant \text{assets} < 1{,}000{,}000, \\ 0 \text{ otherwise}. \end{cases}$$

Suppose $\hat{Y}_R = 1.3 + .7D_1 - .3D_2$. If we plot $\hat{Y}_R$ against the midpoint of assets in each category, we see that the response is not linear (Figure 5.10). The dotted line is speculative.

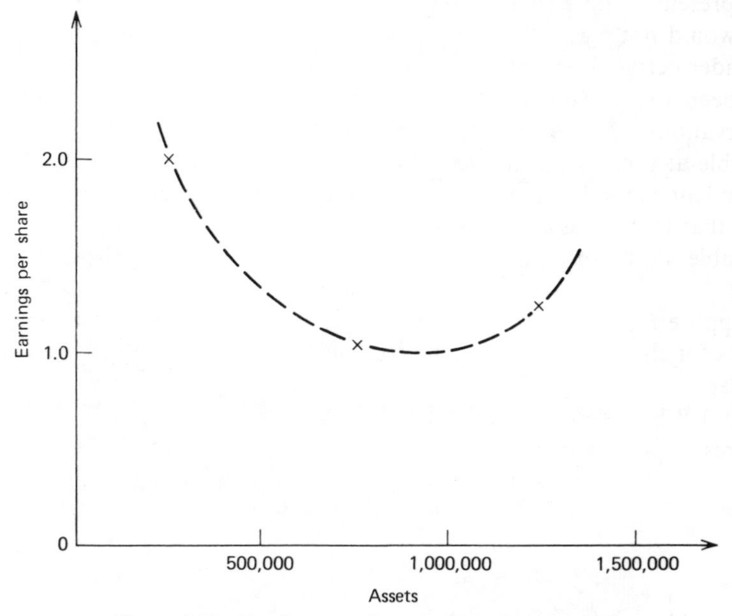

Figure 5.10 Earnings per share as a function of assets.

As usual, there is a price to be paid. By categorizing data, information is destroyed; the fewer the categories, the more the destruction.[1] In addition, one continuous variable is replaced with more than one categorical variable, hence degrees of freedom are lost; this may be serious if the ratio of the number of data points to the number of parameters fitted is small.

[1]The transformation of a continuous variable into dummy variables involves a "many-to-one mapping" in that each dummy variable value represents a range of values in the continuous variable. If this mapping preserves certain characteristics (does not destroy information) to some specified degree, it is called a *homomorphism*. For a good general discussion of transformations and models, see [3], Chapters 6 and 7.

5.5 LACK OF FIT ANALYSIS

A poor fit in terms of a low R^2 and perhaps an F value that is not significant may occur not because of a large error term ϵ but because a linear model was fitted to a nonlinear relationship. The residuals then reflect not only the true error term but also the variation of a true curvilinear relationship from an assumed linear one; for example, in Figure 5.11 the curve Y_R is the true relationship, whereas $\hat{Y}_R$ is a fitted straight line.

The residual *ac* is due to a true error of *bc* and to the component *ab*. The component *ab* would be due mainly to the fact that a straight line was used to represent a curve (note, however, that $\hat{Y}_R$, being a statistic, is a variable and would not match Y_R exactly even if Y_R were a straight line).

Under certain conditions it is possible to test whether the correct model has been fitted. Suppose that there are some data points with identical observations of independent variables. Observations of the dependent variable at such points are called *repeat observations*. The variation of the dependent variable in repeat observations gives a measure for the error term that is free from lack of fit distortion. When repeat observations are available, they can be used to test the null hypothesis that there is no lack of fit.

Suppose i is a set of n_i (where $n_i > 1$) data points, each with the same values for the independent variables. Let there be Q such sets that do not overlap.

Let Y^{ki} be a repeat observation k in the set i. At each set i the sum of squares

$$\sum_{k=1}^{n_i} (Y^{ki} - \bar{Y}^i)^2$$

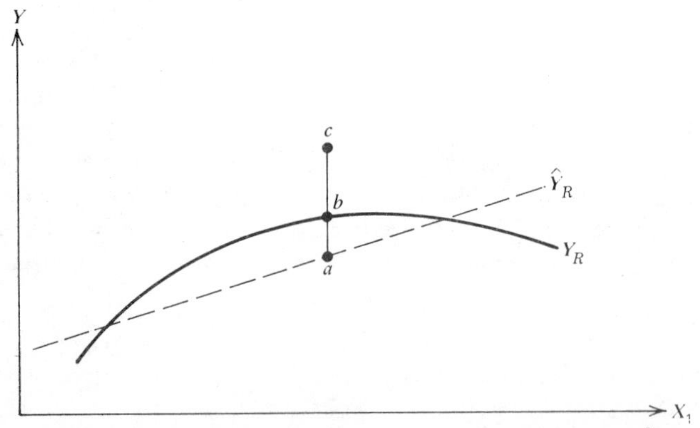

Figure 5.11 A nonlinear relationship Y_R being fitted with $\hat{Y}_R$.

is a measure of the variation in Y from the mean value at i (namely $\bar{Y}^i$) that is caused by the error term only; lack of fit does not enter here. The degrees of freedom of this sum of squares can be found to be $n_i - 1$ by rationalization similar to that used to find the degrees of freedom for the total sum of squares in Table 2.4; that is, one degree of freedom was lost by using $\bar{Y}^i$ which is calculated from the data. For example, suppose that set i consists of the 13'th, 14'th, and 15'th observations:

$$Y_{13} = 7, \quad X_{1,13} = 52, \quad X_{2,13} = 46$$

$$Y_{14} = 8, \quad X_{1,14} = 52, \quad X_{2,14} = 46$$

$$Y_{15} = 9, \quad X_{1,15} = 52, \quad X_{2,15} = 46$$

Because $n_i = 3$ and $\bar{Y}^i = (7 + 8 + 9)/3 = 8$,

$$\sum_{k=1}^{n_i} (Y^{ki} - \bar{Y}^i)^2 = (7-8)^2 + (8-8)^2 + (9-8)^2$$

$$= 2$$

The degrees of freedom are 3 − 1, or 2.

Adding the sums of squares for the Q sets gives

$$\text{sum of squares of pure error} = \sum_{k=1}^{n_1} (Y^{k1} - \bar{Y}^1)^2 + \cdots + \sum_{k=1}^{n_Q} (Y^{kQ} - \bar{Y}^Q)^2 \tag{5.7}$$

The sum of squares of pure error (SS of pure error) is so called because it is "untainted" by any assumption on the form of the relationship being fitted; SS of pure error has n_e degrees of freedom, where

$$n_e = \sum_{i=1}^{Q} (n_i - 1) \tag{5.8}$$

We simply add the degrees of freedom of each of the Q sets of data points with repeat observations. We now define

$$\text{mean square of pure error (MSPE)} = \frac{\text{SS of pure error}}{n_e} \tag{5.9}$$

The mean square of pure error is an unbiased estimator for σ^2, the variance of the error term.

Now consider the sum of squares that is left when the SS of pure error is subtracted from the SS about regression.

$$\text{SS lack of fit} = \text{SS about regression} - \text{SS of pure error} \qquad (5.10)$$

Sum of squares of lack of fit can be considered that part of the SS about regression that is left after compensation for pure error detected in the repeat observations. Now

$$\text{mean square of lack of fit (MSLF)} = \frac{\text{SS lack of fit}}{n - m - n_e} \qquad (5.11)$$

The quantity $n - m - n_e$ is the degrees of freedom and was "obtained" by subtracting the degrees of freedom of the SS of pure error from the degrees of freedom for the SS about regression as in (5.10). The quantity in (5.11) is an estimate for the error variance. However, if there is a lack of fit, it will tend to be biased and will be too large for this purpose.

As a summary of the discussion so far consider Table 5.14 which is a modification of Table 2.4 to allow for pure error.

Table 5.14 Analysis of Variance Table with Pure Error

Source of variation		Sum of squares	Degrees of freedom	Mean square
Due to regression		SS due to regression	$m - 1$	$\frac{\text{SS due to regression}}{m-1}$
About regression	(i)	SS of pure error	n_e	$\frac{\text{SS pure error}}{n_e}$
	(ii)	SS of lack of fit	$n - m - n_e$	$\frac{\text{SS lack of fit}}{n-m-n_e}$
Total sum of squares		SS about the mean	$n - 1$	$\frac{\text{SS about the mean}}{n-1}$

The statistic

$$F = \frac{\text{mean square of lack of fit}}{\text{mean square of pure error}} \qquad (5.12)$$

has an F distribution with $\nu_1 = n - m - n_e$ and $\nu_2 = n_e$ when the null hypothesis that there is no lack of fit is actually true. When there is a lack of fit, the mean square of lack of fit tends to be large and the large value of F that results will cast doubt on the null hypothesis.

▶ **Example 5.7** A researcher performed an experiment in which he compared the performance of male workers on an industrial task thought to require tallness. A "standard" performance meant a score of 100. The researcher had specially trained some of the workers for the task on the theory that training could compensate for lack of height. His experimental results are given in Table 5.15.

Table 5.15 Height, Training, and Task Performance

Worker number	Height in feet (X_1)	Trained $D_2 = 1$ or untrained $D_2 = 0$	Task performance (Y)
1	5.8	0	92
2	5.8	0	96
3	5.8	1	97
4	5.8	1	103
5	5.9	0	115
6	5.9	0	112
7	5.9	0	117
8	5.9	1	118
9	5.9	1	124
10	6.0	0	112
11	6.0	0	116
12	6.0	0	116
13	6.1	1	128
14	6.1	1	117

The regression results are given in Table 5.16. To summarize,

$$\underset{}{Y_R} = \underset{(115)}{-338} + \underset{(19.4)}{75.7X_1} + \underset{(3.97)}{3.42D_2}$$

Table 5.16 Regression Results for the Data on Table 5.15

MULTIPLE R	.77751		F	SIGNIFICANCE
R SQUARE	.60452		8.40731	.006
STD DEVIATION	7.31224			
			SUM OF	MEAN
ANALYSIS OF VARIANCE		DF	SQUARES	SQUARE
REGRESSION		2	899.05739	449.52870
RESIDUAL		11	588.15689	53.46881

Table 5.16 (continued)

VARIABLES IN THE EQUATION

VAR	B	STD ERROR B	F SIGNIFICANCE	BETA ELASTICITY
X_1	75.659824	19.398971	15.211544 .002	.7433829 4.01292
D_2	3.4237537	3.9696854	.74386279 .407	.1643888 .01314
(CONST)	− 337.83871	114.72555	8.6715864 .013	

Because the partial F value of D_2 was not significant, he was led to conclude that training was not important.[1] However, being the cautious sort, he decided to do a lack of fit analysis.

He calculated the pure error sum of squares in Table 5.17.

Following Table 5.14, analysis of variance table (Table 5.18) was prepared.

An F test according to (5.12) gives

$$F = \frac{151.5}{16.7} = 9.07$$

Because $\nu_1 = 3$ and $\nu_2 = 8$, then $F_{.01} = 7.59$. Therefore lack of fit is significant at the .01 level.

The researcher suddenly remembered that he had known all along that the response of task performance to height cannot be linear but must show diminishing returns to that variable. He therefore repeated the regression, adding the variable X_1^2. The result is in Table 5.19.

Because the coefficient of D_2 is now significantly different from 0 (at the .01 level), hindsight has saved the researcher's theory.

Lack of fit analysis is a valuable tool for checking the appropriateness of a model when repeat observations are available. Care should be taken that repeat observations are "genuine." In the preceding example the remeasuring of a worker's performance would not constitute a genuine repeat observation. It would serve only to shed light on the variation in performance for a single worker. The error term in the example included variations *among* workers. ◀

[1]The Durbin-Watson test described in Chapter 6 would have given warning of nonlinearity.

Table 5.17 Sum of Squares Due to Pure Error

Data point set with repeat observations (i)	Observation numbers	Degrees of freedom	Mean $\bar{Y}^i$	Pure error sum of squares
1	1, 2	1	94	$(92-94)^2+(96-94)^2$
2	3, 4	1	100	$(97-100)^2+(103-100)^2$
3	5, 6, 7	2	114.7	$(115-114.7)^2+(112-114.7)^2+(117-114.7)^2$
4	8, 9	1	121	$(118-121)^2+(124-121)^2$
5	10, 11, 12	2	114.7	$(112-114.7)^2+(116-114.7)^2-(116-114.7)^2$
6	13, 14	1	122.5	$(128-122.5)^2+(117-122.5)^2$
Total		8		133.8

Table 5.18 Analysis of Variance Table

Source of Variation	Sum of squares	Degrees of freedom	Mean square
Due to regression	899.1	2	449.5
About regression			
(i) pure error	133.8	8	16.7
(ii) lack of fit	454.4	3	151.5
Total	1487.3	13	114.4

Table 5.19 Regression Equation with X_1^2 Added

			F	SIGNIFICANCE
MULTIPLE R	.92952			
R SQUARE	.86401		21.17786	.000
STD DEVIATION	4.49722			

ANALYSIS OF VARIANCE	DF	SUM OF SQUARES	MEAN SQUARE
REGRESSION	3.	1284.96463	428.32154
RESIDUAL	10.	202.24965	20.22497

VARIABLES IN THE EQUATION

VAR	B	STD ERROR B	F SIGNIFICANCE	BETA ELASTICITY
X_1	7233.0923	1638.5934	19.485233	71.0675334
			.001	383.63618
D_2	8.7251628	2.7264773	10.241022	.4189318
			.009	.03349
X_1^2	− 602.59350	137.95164	19.080737	− 70.3523649
			.001	− 189.30997
(CONST)	− 21587.229	4865.1316	19.688137	
			.001	

EXERCISES

5.1 Refer to the data in Example 5.1. Use a regression program to fit a relationship with Y as the dependent variable and X_1 and $1/X_1$ as independent variables.

(a) Plot the fitted relationship on a diagram that shows the data points.
(b) Which relationship would you prefer: the one fitted in (a) or the one found in Example 5.1 and plotted in Figure 5.1? Why?

5.2 A financial analyst wished to find an estimate for the growth rate of the earnings per share of a common stock. He postulated that earnings per share grew at the rate of $g \times 100\%$ per year, hence fitted

$$\widehat{(\mathrm{EPS})}_R = b_0(1 + g)^t$$

where EPS was the earnings per share at year t and b_0 was an estimator for some parameter β_0. The estimated growth rate g was of primary interest. Data were available for 12 years;

(a) What logarithmic transformation could be used to fit this relationship?
(b) Plot the estimated relationship for years 1 to 12 if $g = .1$ and $b_0 = 2.0$.
(c) In 4 of the 12 years the earnings per share were negative. Discuss the implications for the model used and for the mechanics of the transformation in (a).

5.3 F. B. N. Enterprises was in the business of selling industrial chemicals. It advertised in the trade press and by promotional literature. The company maintained a highly trained sales staff. The president's assistant gathered data for the last six years on the sales, on the number of salesmen on the staff, and on the amount of money spent for advertising. Dollar amounts were adjusted for inflation. Table 5.20 presents the data by quarter.

Table 5.20 Sales, Advertising and Number of Salesmen for 24 Quarters

Observation	Quarter	Sales in thousands of dollars Y	Advertising expenditure in thousands of dollars X_1	Number of salesmen X_2
1	1	5477	10	100
2	2	6805	15	100
3	3	5861	14	103
4	4	5185	12	103
5	1	5024	12	103
6	2	6257	18	103
7	3	6626	16	115
8	4	5712	14	115
9	1	5980	16	115
10	2	5940	18	112
11	3	4752	14	112
12	4	5354	14	112
13	1	6394	15	112
14	2	8674	20	112
15	3	9404	18	125
16	4	10183	17	125
17	1	7459	14	130
18	2	8754	17	130
19	3	11544	24	130
20	4	10843	20	128
21	1	10185	16	127
22	2	10040	16	127
23	3	10855	18	128
24	4	11691	20	128

There were two schools of thought in the company regarding the effect of advertising and number of salesmen on sales. One school held that the dollar gain in sales per dollar increase in advertising was a constant as was the dollar gain in sales per additional salesman. The other opinion held that a 1% change in advertising would always cause a constant percentage change in sales. Roughly a 1% change in sales force would also always cause a constant percentage change in sales.

(a) Construct a model for sales corresponding to each of these schools of thought.
(b) Fit each model with a regression program. Compare and carefully discuss the results.

5.4 A study was done of the market values of homes in two adjacent communities labeled L_1 and L_2. Twenty-seven homes were sold within a week's time. The lot for

Table 5.21 Selling Price, Location, Square Feet, Lot Size, and Quality Index for 27 Homes

House	Selling price ($)	Location	Square feet	Lot size	Quality
1	53,000	L_1	2020	L	3
2	26,000	L_1	980	R	2
3	34,300	L_2	1230	L	3
4	27,000	L_2	980	R	2
5	18,000	L_2	640	R	2
6	18,200	L_2	720	R	2
7	46,000	L_1	2400	L	2
8	18,600	L_2	670	R	1
9	21,350	L_2	690	R	2
10	24,300	L_1	820	R	2
11	39,400	L_1	1910	L	3
12	21,000	L_1	900	R	3
13	32,000	L_1	970	R	3
14	46,000	L_1	1890	R	3
15	21,000	L_2	1050	R	1
16	17,000	L_1	930	R	2
17	20,050	L_1	700	R	2
18	53,500	L_1	1900	L	3
19	42,300	L_1	2100	L	2
20	21,200	L_1	1200	R	2
21	20,500	L_2	680	R	2
22	24,800	L_2	975	R	2
23	17,000	L_2	750	R	1
24	32,000	L_1	1050	R	2
25	18,000	L_2	770	R	1
26	39,000	L_1	2080	L	3
27	53,000	L_1	1900	L	3

each home was labeled R if the lot was of approximately standard size and L if the lot was large. Assume that only two sizes of lot were in the sample. The general quality of construction materials and state of repair was rated 1 if below average, 2 if average, and 3 if above average. The selling price and square feet of space were also recorded. Table 5.21 gives the data.

(a) Construct a linear regression model with selling price as the dependent variable. Use dummy variables for location, lot size, and quality.
(b) Decide on the signs that the coefficients should have.
(c) Run a regression on a computer program and test the significance of each coefficient according to your assumptions in (b). Use the .05 level of significance.
(d) Test the hypothesis that quality has no effect on selling price. Use the .05 level of significance.
(e) Assume that it is believed that spaciousness in top quality homes has a greater effect in influencing prices than in average or below average quality homes. Construct a model in accordance with this assumption and fit the model with a computer program. Evaluate the results.
(f) Consider the use of a selection procedure in this problem. Employ selection with your choice of F_{in} and F_{out} values. Justify your choice of methods and criteria. Have the variables that have been eliminated no effect on selling prices?

5.5 Do a lack of fit analysis on each of the regression relationships found in Exercise 5.3 (b). [See also Exercise 6.2 (d).] Comment.

5.6 Do a lack of fit analysis on the regression relationship found in Exercise 5.4 (c). (*Hint*. Some researchers use points where the independent variable values are nearly identical. If you use this method, you must justify the "nearly.")

5.7 The geographer in Exercise 2.1 did some thinking about the chief causes of long running times. His conclusions were that the time was affected linearly by the number of demand centers but increased as the square of the number of potential sites. Further, because of combinatoric considerations, Y increased with $(X_1/X_2 - .5)^2$.

(a) Construct the new model and run it on a linear regression program.
(b) Perform a lack of fit analysis on this new model and on the linear one in Exercise 2.1. Compare and discuss the results. In the predictive power of which model would you have more confidence?
(c) At one time the geographer had thought that instead of $(X_1/X_2 - .5)^2$ the variable $1/(X_1/X_2 - .5)^2$ should be in the model. He encountered difficulties. Discuss.

REFERENCES

1. Norman Nie, Dale H. Bent and C. Hadlai Hull, *Statistical Package for the Social Sciences (SPSS)*. New York: McGraw-Hill, 1970.
2. N. R. Draper and H. Smith, *Applied Regression Analysis*. New York: Wiley, 1966, Chapter 6.
3. Stafford Beer, *Decision and Control*. New York: Wiley, 1966.

chapter 6

Deviations from the basic regression model

6.1 INTRODUCTION

Our discussion of regression methods and of various related statistical tests and coefficients has, to this point, been based on the linear regression model developed in Section 2.1. To review, it was postulated that an observation of the dependent variable Y_i was "due" to a linear function of independent variables plus a random error term. From (2.2) and (2.3)

$$Y_i = \beta_0 + \beta_1 X_{1i} + \cdots + \beta_q X_{qi} + \epsilon_i \tag{6.1}$$

The error term ϵ, was assumed to be normally distributed with a mean of zero and a constant standard deviation of σ. Furthermore, the ϵ_i's were statistically independent.

It is not reasonable to believe that the model would ever represent reality exactly. The relevant questions, rather, concern the magnitude and types of deviation of the model from reality and the resulting consequences on the validity of regression tests and coefficients.

Key in the study of deviation of the actual data-generating process from the regression model that represents it are the residuals:

$$e_i = Y_i - \hat{Y}_{Ri} \tag{6.2}$$

Residuals were first defined in Section 2.2. The residual is the deviation of the observed value Y_i from the value $\hat{Y}_{Ri}$ predicted by the fitted linear relationship $\hat{Y}_R$. The residual e_i is an estimator for the error ϵ_i; the estimate will generally be good if $\hat{Y}_R$ is close to Y_R.

The residuals can be used to test the original assumptions about normality, constant variance, and statistical independence in the error term. Such tests have implications for the validity of many of the regression statistics commonly turned out by computer programs. They may also point to such modeling errors as missing variables or the false assumption of linearity in the true relationship.

6.2 NONNORMALITY IN THE ERROR TERM

Given that the error term ϵ has constant variance and ϵ_i's are statistically independent, deviations from normality in the distribution of ϵ usually have no serious consequences.

Suppose, for example, that the error term had a nonnormal distribution as in Figure 6.1; what would the consequences be on estimators and statistics printed out by a regression program? As shown in more advanced texts [1], the sample regression coefficients are still unbiased estimators for the regression parameters. The mean square about the regression is still an unbiased estimator for the variance of the error term. These estimators are also consistent (review Section 1.10) and have minimum variance of all possible unbiased linear estimators.

The t and F distributions used for the testing of hypotheses and for interval estimation are no longer strictly correct. In practice, they are still used. These distributions, however, become more appropriate as the sample size is increases. Actually the t tests are called "robust" because they are not very sensitive to the shape of the distribution for ϵ.

Major departures from normality can cause serious errors but unfortunately even major departures from normality are difficult to prove. Apparent nonnormality may in reality be caused by nonlinearity, missing variables, or by a nonconstant error term variance. Bearing this warning in mind, attempts to determine the nature of the error term can be made, either informally or formally.

We could use the methods of descriptive statistics to draw a frequency histogram for the residuals. A bell-shaped distribution would be supportive of the normality assumption. Another approach would be to use *normal probability paper*. When the cumulative percent frequency is plotted for the

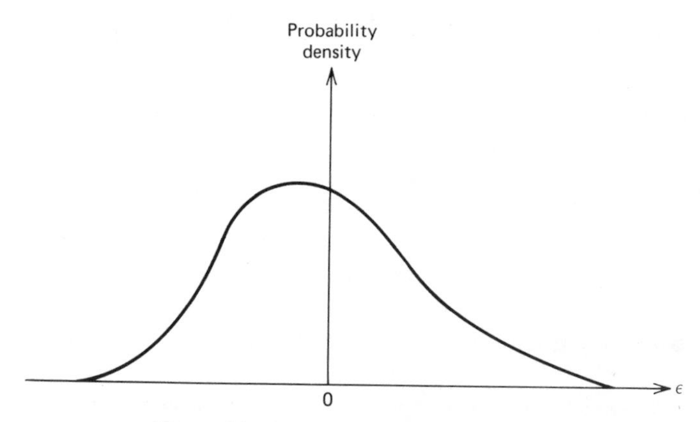

Figure 6.1 A nonnormal distribution for ϵ.

residuals on this paper, it should appear roughly as a straight line. Formal tests such as chi-square goodness-of-fit tests could be used, but this degree of rigor is probably of dubious practical value.

Formal tests for normality are not often done. In large samples lack of normality has no important consequences and in small samples it is difficult to prove.

6.3 HETEROSCEDASTICITY

The rather unpronouncable term *heteroscedasticity* is the formal name for the case in which the error term has no constant variance. *Homoscedasticity* is neatly defined as the absence of heteroscedasticity. The existence of heteroscedasticity is often quite plausible; for example, imagine a study designed to express annual income for individuals as a function of age, education, health, intelligence, and so on. The error term when Y_R, the annual income, is specified by the independent variables as \$60,000 might well have a standard deviation of \$20,000, but it is unlikely that the standard deviation of the error is \$20,000 for a specified income of \$4,000 a year.

The consequences of undetected heteroscedasticity are now summarized. The estimators for the regression parameters can still be shown to be unbiased. These estimators, however, no longer have minimum variance, for if we knew the exact nature of the heteroscedasticity we could construct estimators with smaller variances (see Section 6.4). Furthermore, the standard errors of the sample regression coefficients, given by a regression program, are incorrect (usually the tendency is to underestimate these standard errors). As a result, tests of significance and confidence intervals for the regression coefficients may be seriously misleading.

In its general form heteroscedasticity is impossible to detect because we have only one residual for every observation. One error value could be larger or smaller than another because of randomness and because of a difference in variance of the error term; the effects are impossible to separate.

Detection is feasible only when changes in error term variance are related to another variable. A pattern in the residuals or prior knowledge could then lead one to find heteroscedasticity; for example, consider the residual plot in Figure 6.2. The residuals plotted[1] in Figure 6.2 suggest that the error variance increases as X_j increases. Some computer programs such as the BMD [2] program, have provision for plots of residuals against

[1] Appendix A, Note 2.5, proves that the residuals always have a mean of 0.

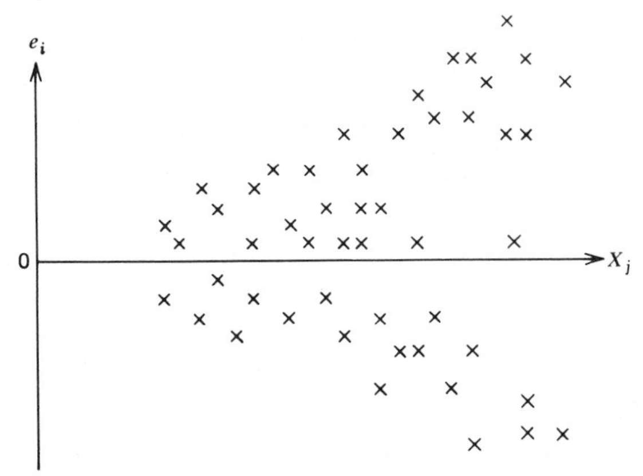

Figure 6.2 Residual pattern suggesting heteroscedasticity.

independent variables; others provide only a plot against the observation number i.

Suppose that a regression with a single independent variable X_1 yielded a residual plot, as in Figure 6.2, and that this plot convinced us that error term variance increases with X_j. It can be proved ([3], 254–256) that the estimated variance of the regression coefficient b_1, in other words $S_{b_1}^2$ (as calculated in the regression), is biased and underestimates $\sigma_{b_1}^2$ on the average. A rather cavalier rationalization for this result is the following:

In Figure 6.3(a) and (b) the scatter of points is represented by a shaded envelope. Assume that in both regressions $\hat{Y}_R$ has the same slope based on the same number of points. Assume also that the mean squared error in both cases is the same. It should seem intuitively appealing that one could

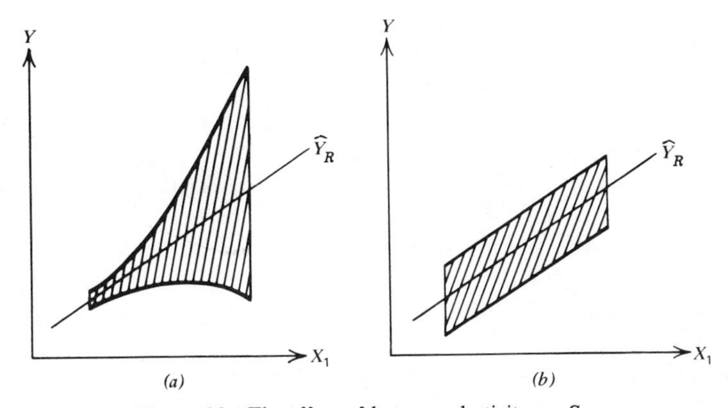

Figure 6.3 The effect of heteroscedasticity on S_b.

be more confident of having found the slope of a line drawn through the envelope of points in Figure 6.3(b) than in the one through the envelope in Figure 6.3(a). The right end of the line $\hat{Y}_R$ in Figure 6.3(a) is not firmly "anchored." The regression program, however, does not "know" that (a) occurred and assumes (b).

6.4 COMPENSATING FOR HETEROSCEDASTICITY

Although the estimators of the regression parameters obtained from straightforward least squares regression can be shown to be unbiased, they are not minimum variance estimators if heteroscedasticity exists. Better estimators could be found. Theoretically, at least, there is no difficulty in finding such estimators.

Suppose that the standard deviation σ_i of the error term is known for each observation i. We can divide (6.1) by σ_i to obtain

$$\frac{Y_i}{\sigma_i} = \frac{\beta_0}{\sigma_i} + \frac{\beta_1 X_{1i}}{\sigma_i} + \cdots + \frac{\beta_q X_{qi}}{\sigma_i} + \frac{\varepsilon_i}{\sigma_i} \tag{6.3}$$

The new error term is ε_i/σ_i. By rule 2 in Section 1.4

$$\begin{aligned} \mathrm{VAR}\left(\frac{\varepsilon_i}{\sigma_i}\right) &= \frac{1}{\sigma_i^2} \cdot \mathrm{VAR}(\varepsilon_i) \\ &= \frac{1}{\sigma_i^2} \cdot \sigma_i^2 \\ &= 1 \end{aligned}$$

Theoretically, then, all we need to do is replace the dependent variable values with Y_i/σ_i and each independent variable value X_{ji} with X_{ji}/σ_i. Unfortunately it is extremely unlikely that the standard deviation for each observation would be known.

Suppose, however, that the examination of residuals reveals a relationship between the standard deviation of the error term and some independent variable X_j; for example, the residuals could form a funnel, the width of which increases with X_j. Suppose that this funnel could be drawn to enclose roughly a constant proportion of residual points at any X_j. If the funnel has straight sides and the distribution of residuals is approximately normal, the standard deviation is roughly proportional to the magnitude of X_j.

Figure 6.4 shows such a funnel that encloses 95% of the points. The width of this funnel is therefore four standard deviations (a property of the

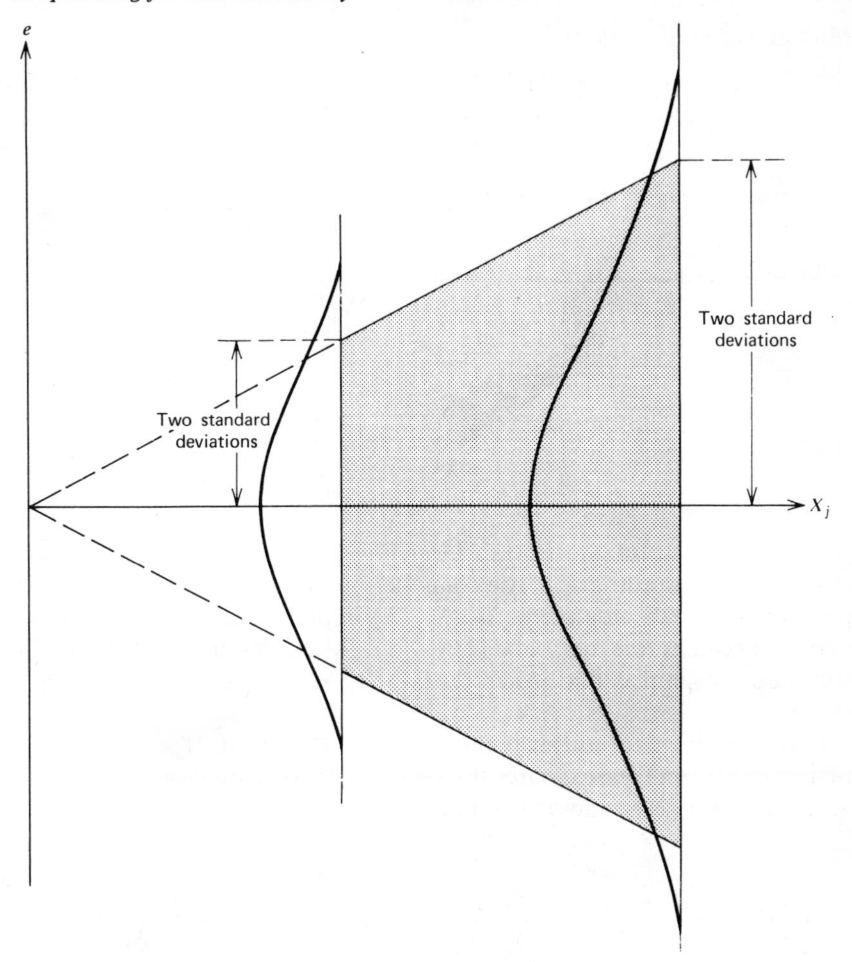

Figure 6.4 The error standard deviation σ_i is proportional to X_j.

normal distribution); hence the standard deviation increases linearly with X_j. Therefore a roughly straight-sided funnel widening to the right implies that σ_i is proportional to X_{ji}.

In this case a transformation can be made that will restore homoscedasticity. Let $\sigma_i = \lambda X_{ji}$, where λ is a constant. Equation 6.3 then becomes

$$\frac{Y_i}{\lambda X_{ji}} = \frac{\beta_0}{\lambda X_{ji}} + \frac{\beta_1 X_{1i}}{\lambda X_{ji}} + \cdots + \frac{\beta_j}{\lambda} + \cdots + \frac{\beta_q X_{qi}}{\lambda X_{ji}} + \frac{\varepsilon_i}{\lambda X_{ji}}$$

Multiplying both sides by λ

$$\frac{Y_i}{X_{ji}} = \frac{\beta_0}{X_{ji}} + \frac{\beta_1 X_{1i}}{X_{ji}} + \cdots + \beta_j + \cdots + \frac{\beta_q X_{qi}}{X_{ji}} + \frac{\varepsilon_i}{X_{ji}} \tag{6.4}$$

The error in term in (6.4) is ε_i / X_{ji} and

$$\begin{aligned} \text{VAR}\left(\frac{\varepsilon_i}{X_{ji}}\right) &= \frac{1}{X_{ji}^2} \cdot \text{VAR}(\varepsilon_i) \\ &= \frac{1}{\sigma_i^2/\lambda^2} \cdot \sigma_i^2 \\ &= \lambda^2 \end{aligned}$$

Because λ^2 is a constant, the transformation in (6.4) is all that is necessary to restore homoscedasticity. The computer program is now given Y_i / X_{ji} in place of Y_i, X_{1i}/X_{ji} in place of X_{1i}, X_{2i}/X_{ji} in place of X_{2i} and so on. Note that β_j becomes the new intercept term. After the transformation the statistical properties that heteroscedasticity destroyed are theroetically restored.

It is interesting to note that the transformation in (6.4) simply implies that in minimizing least squares the residuals of (6.2) are each weighted by $1/X_{ji}$. To fit (6.4) we must minimize[1]

$$\sum_{i=1}^{n} \left[\frac{Y_i}{X_{ji}} - \left(\frac{b_0}{X_{ji}} + b_1 \frac{X_{1i}}{X_{ji}} + \cdots + b_j \frac{X_{ji}}{X_{ji}} + \cdots + b_q \frac{X_{qi}}{X_{ji}} \right) \right]^2$$

$$= \sum_{i=1}^{n} \left(\frac{e_i}{X_{ji}} \right)^2 \tag{6.5}$$

▶ **Example 6.1** A researcher wished to find a linear relationship between the cost of producing a product and the number of units produced. Twenty-two observations are presented in Table 6.1.

The researcher first fitted a relationship $Y_R = \beta_0 + \beta_1 X_1$ and obtained (from Table 6.2)

$$\hat{Y}_R = \underset{(13.2)}{44.4} + \underset{(.00193)}{.0322} X_1 .$$

[1]Review Section 2.2 and compare (6.5) with (2.6) and (2.7).

Table 6.1 Total Cost and Number of Units Produced for 22 Observations

Observation	Cost in thousands of dollars (Y)	Number of units (X)
1	72	1000
2	86	1030
3	84	1050
4	98	1100
5	92	2000
6	156	3050
7	140	3500
8	172	4000
9	215	5000
10	210	5500
11	260	6000
12	210	6500
13	265	7000
14	300	7050
15	320	7450
16	260	8000
17	360	9000
18	310	9500
19	300	9500
20	420	10000
21	370	11100
22	480	12000

Table 6.2 Regression Printout for the Data in Table 6.1

MULTIPLE *R*	.96615	*F*	SIGNIFICANCE
R SQUARE	.93344	280.48787	.000
STD DEVIATION	30.70069		

ANALYSIS OF VARIANCE	DF	SUM OF SQUARES	MEAN SQUARE
REGRESSION	1	264368.81321	264368.81321
RESIDUAL	20	18850.64134	942.53207

VARIABLES IN THE EQUATION

VAR	B	STD ERROR B	F / SIGNIFICANCE	BETA / ELASTICITY
X_1	.32245647E-01	.19253694E-02	280.48787	.9661478
			.000	.81131
(CONST)	44.428399	13.150690	11.413642	
			.003	

The researcher suspected, however, that because cost measurement errors and other error factors increased with total production cost the standard deviation of the error term might be proportional to the number of units produced.

Figure 6.5, a plot of residuals (in units of standard deviations) against observation numbers produced by the SPSS program [4], tends to confirm this hypothesis. In figure 6.5 the first observed residual is plotted at the top and the last at the bottom. Because the observations in Table 6.1 were ordered so that the number of units increased with the observation number, it can be concluded that the standard deviation of the error term is likely to increase with X_1.

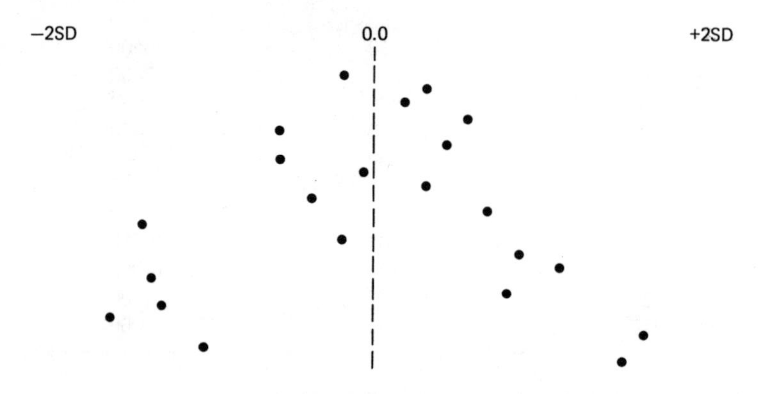

Figure 6.5 SPSS residual printout.

A formal test of the hypothesis that the error variance does not increase with X_1 could be made in the following way. A regression could be run on the first 11 observations and on the last 11 (ordered by increasing X_1). The mean square about the regression is an unbiased estimator for the error variance in each case. The degrees of freedom for each mean square are $11 - 2 = 9$ (review Section 2.4). Then

$$F = \frac{\text{mean square about regression (last 11 observations)}}{\text{mean square about regression (first 11 observations)}}$$

has an F distribution with $\nu_1 = 9$ and $\nu_2 = 9$, provided that the null hypothesis of homoscedasticity is true. A value of F in the critical region for a given level of significance would lead to the rejection of the null hypothesis. Rejection of the null hypothesis, however, would indicate only that the error standard deviation increases with X_1; it would not reveal the exact nature of that relationship.

Figure 6.6 shows a plot of the residuals against values of X_1. A straight-sided funnel (actually a triangle with a vertex at the origin) does not seem totally implausible.

According to (6.4), the researcher decided to fit

$$\left(\frac{Y}{X_1}\right)_R = \beta_1 + \beta_0\left(\frac{1}{X_1}\right)$$

using Y/X_1 as the dependent variable and $1/X_1$ as the independent variable.

Table 6.3 shows that the result is

$$\widehat{\left(\frac{Y}{X_1}\right)}_R = \underset{(.00172)}{.0308} + \underset{(3.84)}{50.9}\left(\frac{1}{X_1}\right)$$

hence $b_0 = 50.9$ and $b_1 = .0308$. The estimated standard errors are smaller.

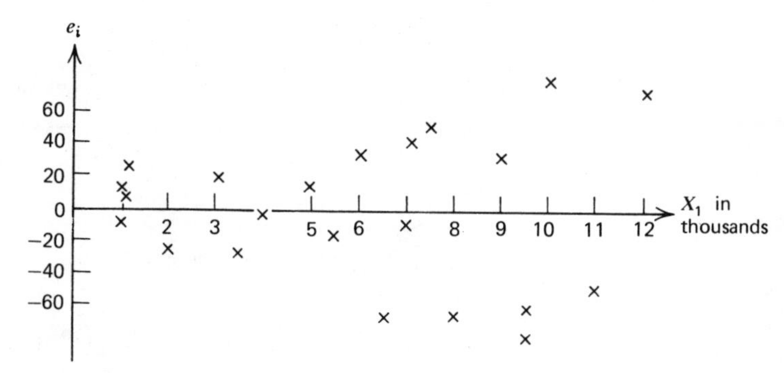

Figure 6.6 Residual plot for the regression of Table 6.2.

Table 6.3 Regression of Y/X_1 Against $1/X_1$

MULTIPLE R	.94760		F	SIGNIFICANCE
R SQUARE	.89794		175.96982	.000
STD DEVIATION	.00567			
			SUM OF	MEAN
ANALYSIS OF VARIANCE		DF	SQUARES	SQUARE
REGRESSION		1	.00566	.00566
RESIDUAL		20	.00064	.00003

Table 6.3 (continued)

VARIABLES IN THE EQUATION

VAR	B	STD ERROR B	F SIGNIFICANCE	BETA ELASTICITY
$\frac{1}{X_1}$	50.913750	3.8380974	175.96982	.9475988
			.000	.34552
(CONST)	.30848676E-01	.172344459E-02	320.38911	
			.000	

Figure 6.7 shows that the residuals no longer have the pronounced funnel shape. The R in Figure 6.7 means that a residual is more than two standard deviations greater than zero. ◀

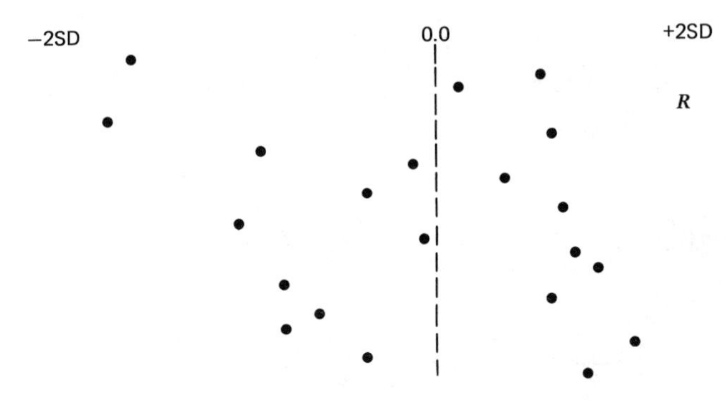

Figure 6.7 SPSS residual plot for the regression of Table 6.3.

An interesting problem develops when it is assumed that σ_i is proportional to $\sqrt{X_{ji}}$ and not to X_{ji} (this assumption is often useful). Equation 6.4 becomes

$$\frac{Y_i}{\sqrt{X_{ji}}} = \frac{\beta_0}{\sqrt{X_{ji}}} + \beta_1 \frac{X_{1i}}{\sqrt{X_{ji}}} + \cdots + \beta_j \sqrt{X_{ji}} + \cdots + \beta_q \frac{X_{qi}}{\sqrt{X_{ji}}} + \frac{\epsilon_i}{\sqrt{X_{ji}}} \tag{6.6}$$

The error term can easily be shown to be constant. Equation 6.6, however, now has no intercept term! This implies that the hyperplane must pass through the origin. To fit (6.6), therefore, *forcing through the origin* must be employed. Theoretically this is no problem because least squares are simply minimized for a relationship without a constant term.[1] Many computer programs allow regression through the origin as an option.

▶ **Example 6.2** Suppose that the researcher in Example 6.1 believed that the standard deviation of the error term was proportional to $\sqrt{X_1}$ and not to X_1. He then fitted

$$\left(\frac{Y}{\sqrt{X_1}}\right)_R = \frac{\beta_0}{\sqrt{X_1}} + \beta_1\sqrt{X_1}$$

Using the SPSS regression through the origin option, (Table 6.4), he found

$$\left(\widehat{\frac{Y}{\sqrt{X_1}}}\right)_R = \underset{(6.33)}{48.9}\ \frac{1}{\sqrt{X_1}} + \underset{(.00147)}{.0315}\ \sqrt{X_1}$$

Table 6.4 Regression of $Y/\sqrt{X_1}$ Against $1/\sqrt{X_1}$ and $\sqrt{X_1}$

			F	SIGNIFICANCE
MULTIPLE R	.99384			
R SQUARE	.98771		803.67679	.000
STD DEVIATION	.36474			

ANALYSIS OF VARIANCE	DF	SUM OF SQUARES	MEAN SQUARE
REGRESSION	2	213.83217	106.91609
RESIDUAL	20	2.66067	.13303

VARIABLES IN THE EQUATION

VAR	B	STD ERROR B	F / SIGNIFICANCE	BETA / ELASTICITY
$\frac{1}{\sqrt{X_1}}$	48.913194	6.3266883	59.772189	.2788725
			.000	.25697
$\sqrt{X_1}$	.31488604E-01	.14701300E-02	458.77036	.7725984
			.000	.74427
(CONST)	0.	0		

The pattern of residuals in Figure 6.8 does not show evidence of heteroscedasticity. Note that the coefficient of determination R^2 is much higher

[1]The mathematical derivation in Appendix A, Note 2.2, could be modified to exclude β_0.

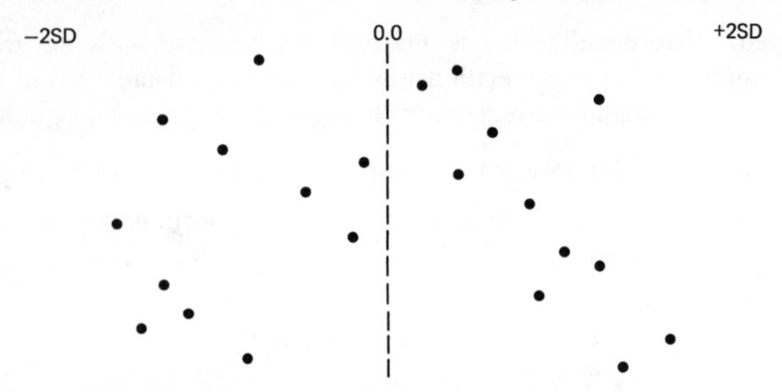

Figure 6.8 SPSS residual plot for the regression of Table 6.4.

in Table 6.4 than in Table 6.3. Comparing R^2 values when transformations are involved can be misleading because different criteria for a "best" fit are employed. ◀

This section concludes with some remarks in summary. First, correcting for heteroscedasticity would be a straightforward procedure if only all the different error variances were known. Such knowledge is highly improbable and, at best, we can link the error variance with some single independent variable.

Such a link is best established on a priori grounds, that is, through knowledge of the process being studied. Deduction of a link by study of the pattern of residuals can be dangerous, especially if the regression sample is small. Spurious patterns are notoriously easy to discern if one is looking for patterns.

6.5 AUTOCORRELATION

One of the original assumptions of the regression model was that successive error term values were uncorrelated with any previous values. *Autocorrelation*, or *serial correlation*, is the statistical dependence of errors on preceding errors. This condition occurs most often in data taken from observations over time. It may be due to undiscovered nonlinearity in the regression relationship or to missing variables. *Positive autocorrelation* is said to exist when the correlation in question has a positive sign.

In this section we will see that if positive autocorrelation is not compensated for, the standard errors of the sample regression coefficients, printed out by a regression program, may underestimate the true standard deviations of the sample regression coefficients ($S_{b_i}^2$ is a biased estimator for $\sigma_{b_i}^2$). The R^2 value could also be higher than it should be. Although the parameter estimators are still unbiased, they will no longer be the minimum variance estimators; better estimators can be found.

Let us examine first some examples of how positive autocorrelation may be caused. In Figure 6.9 a straight line was fitted to what appears to be a nonlinear relationship. Note that there are groups of negative and positive residuals. Figure 6.10 shows a plot of adjacent residuals for Figure 6.9. (Assume that X_1 increases with the data point number i.) There is evidence of a linear relationship; hence succeeding residual observations are likely to be correlated (review Section 1.13).

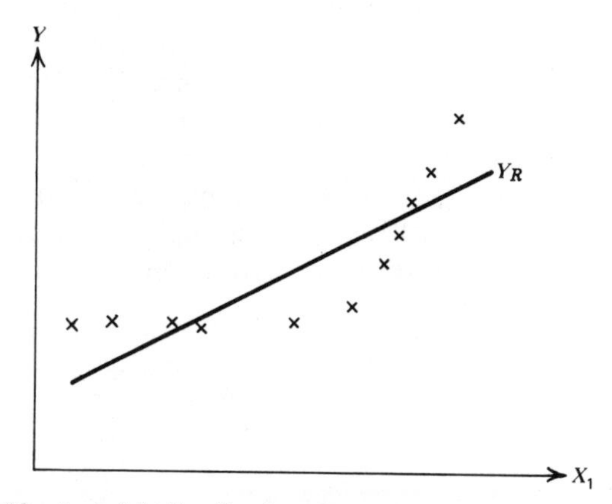

Figure 6.9 A straight line fitted to an apparently nonlinear relationship.

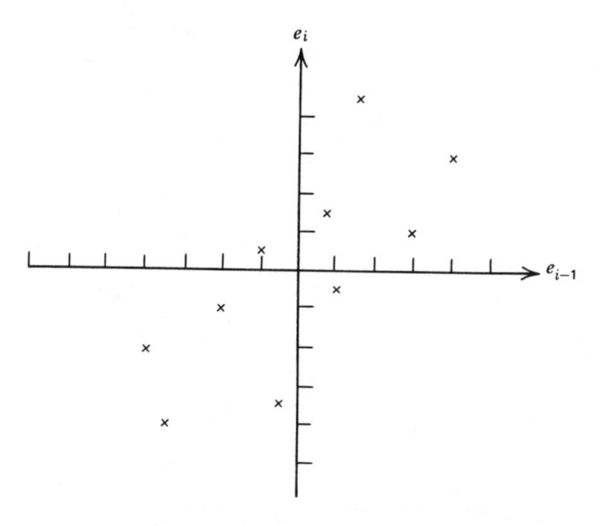

Figure 6.10 Plot of e_i against e_{i-1}.

An example of how a missing variable may cause autocorrelation is shown in Figure 6.11. Suppose that Y is sales, X_1 is advertising, and a linear response of sales to advertising exists. Suppose, however, that per capita disposable income X_2 has been omitted from the model but that increases in disposable income shift the response line upward. Advertising increased over time and so did disposable income. Plotted against X_1 only, the data points seem to form a nonlinear relationship; hence autocorrelation is present if a straight line is fitted.

Still another reason for autocorrelation may exist in the nature of the error term. The error term may be due to causes that act "randomly" but act over more than one period; two or more successive observations may record the same disturbance, hence may be correlated.

Let us now turn to the statistical consequences of autocorrelation. An analytic treatment is beyond the scope of this book, but some possibilities are explained graphically.

Consider Figures 6.12*a* and *b*. Assume that the data points in each diagram were generated by a linear relationship Y_R and by an error term with high positive serial correlation. A different starting point for the

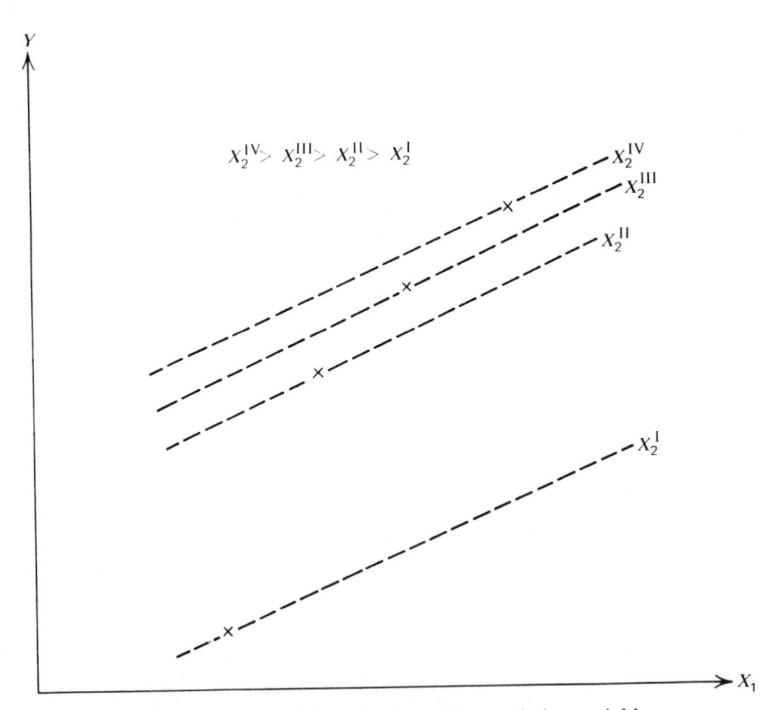

Figure 6.11 Nonlinearity caused by a missing variable.

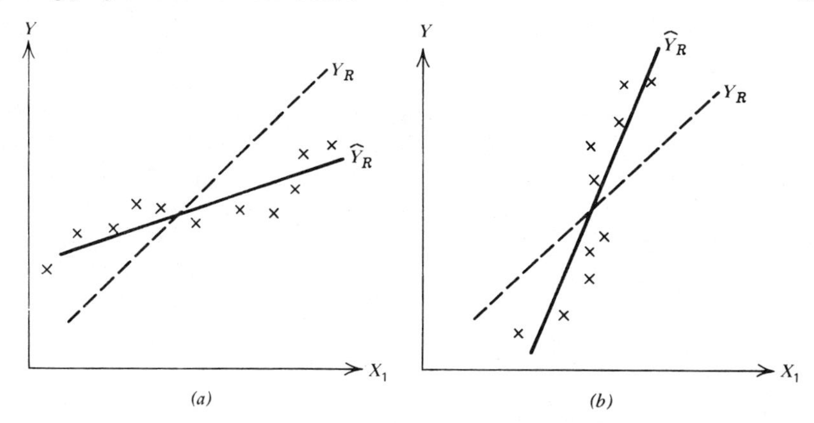

Figure 6.12 The effect of autocorrelation on the slope of a fitted line.

residuals in each case has resulted in the two patterns shown. Note that the regression lines in each case fit the data better than the true line in the sense that the sum of squared residuals will be less than the sum of squared errors.

As a result the calculated R^2 will be higher than it should be and the standard errors of the sample regression coefficients will tend to be underestimated by printout values. In the cases discussed we would obtain a lower S_{b_1} than justified. The parameter estimators b_0 and b_1, however, will be unbiased because no systematic errors are introduced (in the long run estimates that are too high will compensate for estimates that are too low).

The illustrations given above showed only the most common type of autocorrelation: positive autocorrelation of adjacent residuals. Positive residuals tended to follow positive residuals and negative residuals tended to follow negative residuals. In fact, autocorrelation can occur in more complex ways ([5], Chapter 8), but only positive correlation of adjacent terms is dealt with in this book.

6.6 TESTING FOR FIRST ORDER AUTOCORRELATION

A simple model of a positive autocorrelation effect will enable us to study methods for measuring autocorrelation and to devise a way of reducing its disadvantages in obtaining parameter estimators.

Consider

$$Y_i = \beta_0 + \beta_1 X_{1i} + \cdots + \beta_q X_{qi} + \epsilon_i \tag{6.7}$$

where

$$\epsilon_i = \rho\epsilon_{i-1} + \nu_i.$$

The variables ν_i are assumed to be statistically independent, normally distributed, and have constant variance σ_ν^2. It can be shown[1] that if the absolute value of ρ is less than one ($|\rho| < 1$), then ρ represents the correlation coefficient of ϵ_i and ϵ_{i-1}; ρ is called the *coefficient of autocorrelation*. The error term ϵ_i will have its mean equal to zero and variance equal to $\sigma_\nu^2/(1 - \rho^2)$. Because autocorrelation is usually found in time series data, some texts use the subscript t instead of i. This model is called a *first-order autocorrelation model* because correlation is specified only between adjacent error terms.

It is useful to obtain some statistic for estimating ρ. Such an estimator exists and theoretically could be used to test the null hypothesis that $\rho = 0$. In fact, it is used mainly in attempts to compensate for the effects of autocorrelation.

To estimate ρ a linear relationship is fitted between successive residuals.

$$\widehat{(e_i)}_R = re_{i-1} \tag{6.8}$$

The estimator r can be found by least squares techniques[2]; note that we are fitting a line through the origin.

$$r = \frac{\sum_{i=2}^{n} e_{i-1}e_i}{\sum_{i=2}^{n} e_{i-1}^2} \tag{6.9}$$

As an example of the calculation of r consider the residuals $e_1 = -12$, $e_2 = 12$, $e_3 = 3$, $e_4 = 20$, $e_5 = -17$, $e_6 = -6$.

$$\begin{aligned} r &= \frac{e_1e_2 + e_2e_3 + e_3e_4 + e_4e_5 + e_5e_6}{e_1^2 + e_2^2 + e_3^2 + e_4^2 + e_5^2} \\ &= \frac{-12(12) + 12(3) + 3(20) + 20(-17) + (-17)(-6)}{(-12)^2 + 12^2 + (3)^2 + 20^2 + (-17)^2} \\ &= -.29 \end{aligned}$$

It would be useful if r could be used to test the hypothesis that $\rho = 0$. Unfortunately the statistic r has a complex distribution when $\rho = 0$.

[1] See Appendix A, Note 6.1. If $\rho > 1$, ϵ_i would increase without limit.
[2] See Appendix A, Note 6.2.

The test of the hypothesis H_0: $\rho = 0$ against H_1: $\rho > 0$ is often done with the *Durbin-Watson statistic D*, where

$$D = \frac{\sum_{i=2}^{n} (e_i - e_{i-1})^2}{\sum_{i=1}^{n} e_i^2} \tag{6.10}$$

A small value of D would indicate that successive residuals tend to be close together; hence it would make us doubt H_0. This can be seen in another way by means of a useful approximate relationship[1]:

$$D \approx 2(1 - r) \tag{6.11}$$

For $r \approx +1$, D approaches 0, for small positive values of r it approaches 2, and for negative values of r it is greater than 2. Therefore small values of D should lead to rejection of H_0.

The distribution of D is also complex because it can be shown to depend on the particular observed values of the independent variables in the sample. However, limits on its position and shape can be determined independently of the X_j's. This leads to a rather curious decision rule for the rejection of H_0. For a given significance level positive numbers d_U and d_L are given such that $d_U \geqslant d_L$.

Rule: If $D > d_U$, fail to reject H_0 (no autocorrelation); if $D < d_L$, reject H_0; if $d_L \leqslant D \leqslant d_U$, the test is inconclusive. (6.12)

The no-man's-land between d_L and d_U exists because the exact distribution for D is not available. Table 4 in Appendix C lists values of d_L and d_U at the .05 and .01 significance levels; these values are, of course, different for different values of n and q. When the test is inconclusive, more data should be used; when this is not feasible, other, more advanced methods for testing autocorrelation can be found in the statistics literature ([5], Chapter 8).

▶ **Example 6.3** An economist in the country of Lemuria was studying the response of the quantity of beef consumed to the price of beef in dollars and to the per capita annual disposable personal income in hundreds of dollars. Eighteen years of data are presented sequentially in Table 6.5.

[1]See Appendix A, Note 6.3.

Table 6.5 Demand for Beef, Price of Beef, and Disposable Personal Income for 18 Years

Year	Quantity of beef consumed (Y)	Price of beef (X_1)	Disposable personal income (X_2)
1	5642	.90	29
2	5029	.93	30
3	4331	.97	30
4	4456	.99	32
5	5036	.97	34
6	4598	.96	36
7	5036	.90	40
8	6024	.85	47
9	6012	.87	55
10	7029	.89	62
11	6905	.93	63
12	7455	.95	71
13	7584	.95	60
14	7135	.97	60
15	7153	.99	71
16	6877	1.00	73
17	5443	1.18	77
18	6024	1.15	77

The results of the linear regression presented in Table 6.6 give the equation

$$\begin{aligned}\hat{Y}_R = {} & \underset{(1466)}{9880} \; - \underset{(1678)}{7747X_1} \; + \underset{(7.975)}{62.45X_2}\end{aligned}$$

The residual plot (Figure 6.13) shows evidence of positive autocorrelation.

The Durbin-Watson statistic printed was .779.

In Table 4 (Appendix C) for $n = 18$ and $q = 2$, $d_L = 1.05$ and $d_U = 1.53$ at the .05 significance level; $d_L = .80$ and $d_U = 1.26$ at the .01 level of significance. The null hypothesis that there is no positive autocorrelation is rejected at the .01 level of significance.

Using (6.11), we have

$$r \approx 1 - \frac{D}{2}$$
$$= .61$$

An exact value of r could be calculated from the residuals ($r = .557$). ◀

Table 6.6 Regression Printout for the Data in Table 6.5

			F	SIGNIFICANCE
MULTIPLE *R*	.89714			
R SQUARE	.80486		30.93489	.000
STD DEVIATION	513.43405			
			SUM OF	MEAN
ANALYSIS OF VARIANCE		DF	SQUARES	SQUARE
REGRESSION		2	16309770.57485	8154885.28742
RESIDUAL		15	3955217.92516	263614.52834

VARIABLES IN THE EQUATION

VAR	*B*	STD ERROR *B*	*F* / SIGNIFICANCE	BETA / ELASTICITY
X_1	− 7447.0882	1677.8808	19.699299	−.5783743
			.000	− 1.19893
X_2	62.448782	7.9751029	61.316218	1.0204020
			.000	.54876
(CONST)	9879.8325	1466.2651	45.401917	
			.000	

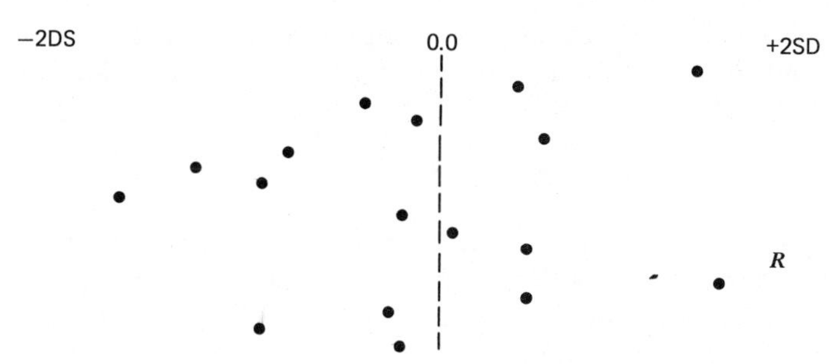

Figure 6.13 SPSS residual plot for the regression of Table 6.6.

6.7 COMPENSATING FOR FIRST-ORDER AUTOCORRELATION

When the Durbin-Watson test indicates autocorrelation, we should consider first the possibility of nonlinearity and also check for missing variables. If the difficulty cannot be eliminated in this way, the problems in parameter estimation caused by autocorrelation can be dealt with otherwise.

Using (6.7), we have

$$Y_{i-1} = \beta_0 + \beta_1 X_{1,i-1} + \cdots + \beta_q X_{q,i-1} + \epsilon_{i-1}$$

Multiplying the above by ρ and subtracting from (6.7), we obtain

$$Y_i - \rho Y_{i-1} = \beta_0(1-\rho) + \beta_1(X_{1i} - \rho X_{1,i-1}) + \cdots$$
$$+ \beta_q(X_{qi} - \rho X_{q,i-1}) + \epsilon_i - \rho\epsilon_{i-1} \qquad (6.13)$$

The new error term $\epsilon_i - \rho\epsilon_{i-1}$ is equal to ν_i; hence there is no correlation. If ρ were known, we could use $Y_i - \rho Y_{i-1}$ as the new dependent variable and $X_{ji} - \rho X_{j,i-1}$ as each independent variable to obtain estimators for $\beta_0(1-\rho), \beta_1, \ldots, \beta_q$. All the desirable statistical properties of an ordinary least squares linear regression would be restored.

Unfortunately ρ is generally not known. The statistic r is often used. However, r is actually calculated under the assumption that $\rho = 0$. If ρ is not zero, the estimates for the regression parameters [obtained by estimating (6.13)] are different. Hence r should be estimated simultaneously with an estimate of the regression parameters. Two approximation methods are discussed.

In the iterative approach r is first calculated by using (6.9); r is then substituted for ρ in (6.13) and new estimates for $\beta_0, \beta_1, \ldots, \beta_q$ are found. The residuals of this run are tested with the Durbin-Watson statistic. If autocorrelation is still judged to be present, the process continues: the new estimates $b_0, b_1, \ldots, b_q$ are used to obtain new residuals [in the relationship estimated for (6.1)]. The residuals are used to calculate a new r through (6.9). Equation 6.13 is fitted and tested for autocorrelation. The iterations are stopped when the hypothesis of zero autocorrelation cannot be rejected.

▶ **Example 6.4** Suppose that the economist in Example 6.3 decided to compensate for autocorrelation and to obtain new regression coefficients through the iterative approach. His r was .557 (or .55685 rounded). According to (6.13) he estimated the relationship

$$(Y_i - .557Y_{i-1})_R = \beta_0(1 - .557) + \beta_1(X_{1i} - .557X_{1,i-1})$$
$$+ \beta_2(X_{2i} - .557X_{2,i-1})$$

A regression was run with the data in Table 6.7. Computer programs usually have internal transformations that create "lagged" variables like Y_{i-1}, making manual calculation of a table like Table 6.7 unnecessary in practice.

$$\widehat{(Y_i - .557Y_{i-1})_R} = \underset{(610.1)}{5100} - \underset{(1574)}{9865}(X_{1i} - .557X_{1,i-1})$$
$$+ \underset{(12.05)}{83.13}(X_{2i} - .557X_{2,i-1})$$

where

$$R^2 = .8078$$
$$F = 29.42$$
$$\text{Durbin-Watson statistic} = 2.60$$
$$\text{SS due to regression} = 6141930 \quad \text{df} = 2$$
$$\text{SS about regression} = 1461230 \quad \text{df} = 14$$

Table 6.7 The Data in Table 6.5 Adapted for Example 6.4

	Quantity of beef consumed		Price of beef		Disposable personal income	
Year	Y_i	Y_{i-1}	X_{1i}	$X_{1,i-1}$	X_{2i}	$X_{2,i-1}$
2	5029	5642	.93	.90	30	29
3	4331	5029	.97	.93	30	30
4	4456	4331	.99	.97	32	30
5	5036	4456	.97	.99	34	32
6	4598	5036	.96	.97	36	34
7	5036	4598	.90	.96	40	36
8	6024	5036	.85	.90	47	40
9	6012	6024	.87	.85	55	47
10	7029	6012	.89	.87	62	55
11	6905	7029	.93	.89	63	62
12	7455	6905	.95	.93	71	63
13	7584	7455	.95	.95	60	71
14	7135	7584	.97	.95	60	60
15	7153	7135	.99	.97	71	60
16	6877	7153	1.00	.99	73	71
17	5443	6877	1.18	1.00	77	73
18	6024	5443	1.15	1.18	77	77

For the .05 level of significance Table 4 in Appendix C ($n = 17$, $q = 2$) lists $d_L = 1.02$ and $d_U = 1.54$. The null hypothesis of no autocorrelation cannot be rejected at this level and the iterative process stops.

The final relationship is

$$\hat{Y}_R = \frac{5100}{1 - .557} - 9865X_1 + 83.13X_2$$

or

$$\hat{Y}_R = \underset{(1377)}{11512} - \underset{(1574)}{9865X_1} + \underset{(12.05)}{83.13X_2}$$

Note that the sample standard error of b_0 was calculated as 610.1/(1 − .557). This equation would have been used to calculate new residuals, hence a new r, if the iterative process had been required to continue. ◀

A simpler alternative to the iterative approach is the two-step Durbin procedure [6]. Equation 6.13 is rewritten slightly to give

$$Y_i = \beta_0(1-\rho) + \rho Y_{i-1} + \beta_1 X_{1i} - \rho\beta_1 X_{1,i-1} + \cdots + \beta_q X_{qi} - \rho\beta_q X_{q,i-1} + \epsilon_i - \rho\epsilon_{i-1} \tag{6.14}$$

This equation will produce an estimate of ρ directly as a coefficient of Y_{i-1}. In a sense r is now obtained "simultaneously" with the estimated regression coefficients. However, only the r is used.[1] Substitution of r in (6.13) and the resulting regression produces new estimates of the regression coefficients.

▶ **Example 6.5** The economist in Example 6.3 decided to compensate for positive autocorrelation with the Durbin procedure. Using the data in Table 6.7, he found

$$\widehat{(Y_i)}_R = \underset{(2119)}{5660} + \underset{(.1784)}{.5012}\,Y_{i-1} - \underset{(1797)}{9544}\,X_{1i} + \underset{(2574)}{4641}\,X_{1,i-1} + \underset{(18.32)}{21.66}\,X_2 + \underset{(23.89)}{19.35}\,X_{2,i-1}$$

Using $r = .5012$, he then found

$$\left(\widehat{Y_i - .5012Y_{i-1}}\right)_R = \underset{(654.3)}{5715} - \underset{(1508)}{9737}\,(X_{1i} - .5012X_{1,i-1}) + \underset{(10.61)}{80.87}\,(X_{2i} - .5012X_{2,i-1})$$

where

$$R^2 = .8288$$

$$F = 33.88$$

Durbin-Watson statistic = 2.59

SS due to regression = 6871340 df = 2

SS about regression = 1419620 df = 14

[1]Note that according to (6.14) estimates of, say, ρ, β_1, *and* $\rho\beta_1$ would be provided and that the estimate for $\rho\beta_1$ might not equal rb_1. The imposition of constraints like $\rho\beta_1 = rb_1$ is possible but not with ordinary linear regression packages.

The final relationship was

$$\hat{Y}_R = \underset{(1312)}{11458} - \underset{(1508)}{9737}X_1 + \underset{(10.61)}{80.87}$$

which is not very different from the one found in Example 6.4. ◀

Some researchers use the *first difference approach* by assuming $\rho = 1$ in (6.13). It is possible to overcorrect for positive autocorrelation in this way and to produce large negative autocorrelation. Note also that by assuming $\rho = 1$ the constant term is eliminated from (6.13); hence a "regression through the origin" option[1, 2] must be used with the program.

6.8 OTHER PROBLEMS

This chapter concludes with a brief selection of additional problems that may occur in fitting the linear regression model. A detailed study of these problems is the proper province of more advanced texts, one that requires more sophistication in statistical techniques. Also, these problems generally require facilities in addition to simple regression programs, hence are beyond the mandate of this book.

The correlation of the error term ϵ with some independent variable X_j is difficult to detect because the nature of the least squares derivation guarantees that the residuals are statistically independent of each X_{ji}. If such a correlation of the error term to an independent variable exists, it can be shown to cause both bias and inconsistency in the parameter estimators. This condition most commonly appears when the data must be explained not by one linear relationship but by two or more acting simultaneously. This *simultaneous equations problem* is treated in most econometrics books [5].

In some applications researchers wish to relate values of the independent variable to past observations of the dependent variable; for example, Y_i may appear as the dependent variable and Y_{i-1}, as the independent variable. These models are called *autoregressive* or *lagged variables* and can be shown to cause independent variables that are correlated to preceding error terms. Parameter estimators become biased.

Applications also exist in which the researcher may wish to use a dummy variable (0 or 1) for the dependent variable ([7], pp. 326–339); the researcher may also often wish to interpret his regression as predicting the

[1]See the discussion following (6.6).

[2]The Durbin-Watson test is not strictly valid for regressions forced through the origin.

probability of Y occurring. Among the statistical problems thus created are heteroscedasticity and nonnormality of error terms. Attempts at correcting the heteroscedasticity follow the general principles established in Section 6.4.

EXERCISES

6.1 Refer to Exercise 5.4 (c).

(a) Find the proportion of residuals within .5 standard deviations from their mean, 1 standard deviation from their mean, and 2 standard deviations from their mean. Are the results compatible with the hypothesis that the residuals have an approximately normal distribution? If your regression program has the capability, obtain a histogram of the residuals.
(b) Plot the residuals against the "space-in-square-feet" variable. Do you suspect that heteroscedasticity is present? Perform a formal test for heteroscedasticity. Investigate the other independent variables for their association with heteroscedasticity.
(c) Assuming that the error term standard deviation is proportional to square feet, run the regression with the proper transformation and again plot the residuals against square feet. Comment on changes in various regression statistics.
(d) Repeat (c), assuming that error term *variance* is proportional to square feet of space.

6.2 Refer to Exercise 5.3.

(a) Plot the residuals for each of the models in Exercise 5.3 (b) against the observation number.
(b) Obtain the Durbin-Watson statistic for each model in Exercise 5.3 (b). For each hypothesis determine that there is no autocorrelation at the .01 level of significance.
(c) Obtain estimates for the autocorrelation coefficients, using both (6.9) and the approximation from the Durbin-Watson statistic.
(d) Review the results of Exercise 5.5. Was the lack of fit test valid? Explain.

6.3 Using the results of Exercise 6.2 (c), try to compensate for autocorrelation in the linear model by

(a) the iterative approach;
(b) the Durbin method. For each method obtain new estimates for the coefficients and for the standard errors of the coefficients. Discuss the results.

6.4 A researcher felt that the easiest way to compensate for autocorrelation detected by tests is to arrange the observations at random and then rerun the regression. Because successive residuals would then be uncorrelated, he felt that the problem would be solved. Comment.

REFERENCES

1. E. Malinvand, *Statistical Methods of Econometrics*. Chicago: Rand McNally, 1966.
2. *Biomedical Computer Programs*, Health Sciences Computing Facility, Department of Biomathematics, School of Medicine, University of Los Angeles, University of California Press, January 1973.
3. Jan Kmenta, *Elements of Econometrics*. New York: Macmillan, 1971.
4. Norman Nie, Dale H. Bent, and C. Hadlai Hull, *Statistical Package for the Social Sciences (SPSS)*. New York: McGraw-Hill, 1970.
5. J. Johnston, *Econometric Methods*. New York: McGraw-Hill, 1963.
6. J. Durbin, "Estimation of Parameters in Time-Series Regression Models," *Journal of the Royal Statistical Society*, Series B, **22**, 139–153, January 1960.
7. J. Neter and W. Wasserman, *Applied Linear Statistical Models*. Homewood, Ill.: Irwin, 1974.

chapter 7

Regression and single-factor analysis of variance

7.1 INTRODUCTION

Analysis of variance, or *ANOVA*, models are statistical models in which, generally, a quantitative dependent variable is influenced by qualitative[1] independent variables acting within a linear relationship. The additive error term is usually assumed to have a normal distribution with an expected value of 0 and a constant variance. The reader may wonder what the difference is between this model and the linear regression model with dummy variables, introduced in Section 5.4. There is no difference. ANOVA models of the "fixed effects" type can be analyzed by regression.

There are reasons why ANOVA models are often treated separately from regression topics. First of all, ANOVA models are generally analyzed by a partitioning of sums of squares procedure that is computationally more efficient than regression, although it yields identical results. This procedure, which, incidentally, has given these models their name, is made possible by the rather fortuitous structure of ANOVA models. Second, ANOVA models have been historically associated with experimental studies in which the independent variables are strictly controlled for the sample under study. As a result a special terminology has been developed for ANOVA models.

The purpose of this chapter, and of the next, is to apply the dummy variable regression approach developed in Section 5.4 to ANOVA models of the fixed effects type. This does not mean that the reader will be restricted to regression programs for the analysis of such models. The interpretation of program outputs in which the results were internally obtained by the partitioning of sums of squares is illustrated. For purposes of interpretation the actual mechanism of computation within the program is irrelevant.

[1]Quantitative independent variables are used in what is called the analysis of covariance, discussed in Chapter 8.

Also, although the regression approach may often be computationally inferior to the partitioning approach, it has one great practical advantage. Sometimes it is difficult to find the proper ANOVA program for the data being analyzed, especially if the model is nonstandard. Even standard experimental designs may become nonstandard if data are missing or invalidated. The regression approach is always applicable and, when efficient programs are available, often practical.

It should be stressed that the emphasis in the following two chapters is not on the many considerations in achieving a good experimental design. Instead, it is on developing a unified method of analysis. In addition, these chapters provide a convenient vehicle for elaboration on the concept of simultaneous statistical inference, a topic of considerable importance in ANOVA models and applicable to other types of regression analysis as well. Chapter 7 develops a basic methodology and terminology; in a sense, the widely used models of Chapter 8 are merely examples of these methods.

7.2 AN ILLUSTRATION

Before introducing the special form and terminology of the single-factor ANOVA model we shall first consider an example problem and its analysis by straightforward linear regression, using dummy variables.

▶ **Example 7.1** An experiment was performed to determine the effect of three different bag designs on the sales of potatoes in supermarkets. Design 1 was radically new; design 2 was similar to those in use by competitors; and design 3 was currently in general use. Eighteen stores, judged to be as much of the same type and size as possible, were chosen for the experiment. By the use of a table of random numbers eight stores were assigned to test design 1 and five stores each to designs 2 and 3. Design 1 was given more stores because it was of particular interest. Price, quality, competition, and other variables remained the same during the week of the test. The sales that resulted are given in Table 7.1.

Suppose that the following model is assumed for the process that generated the data in Table 7.1. Designs 1, 2, and 3 produce different average sales. Deviations from each average are due to random differences in stores, and are normally distributed with a mean of zero and a constant (for all three designs) standard deviation.

This model can be expressed as a linear regression model:

$$Y = \beta_0 + \beta_1 X_1 + \beta_2 X_2 + \epsilon$$

Table 7.1 Sales in Number of Bags per Week

Design 1	Design 2	Design 3
52	23	21
36	36	32
24	29	17
48	16	24
52	27	15
27		
48		
41		

where Y = sales per week,
X_1 = 1 if design 1 was used and 0 otherwise,
X_2 = 1 if design 2 was used and 0 otherwise.

Note that for design 1, $Y = \beta_0 + \beta_1 + \epsilon$; for design 2, $Y = \beta_0 + \beta_2 + \epsilon$; for design 3, $Y = \beta_0 + \epsilon$. Because β_0, β_1, and β_2 are constants and ϵ has a mean of zero, the average sales for design 1 is $\beta_0 + \beta_1$, for design 2, $\beta_0 + \beta_2$, and for design 3, β_0. The error term ϵ is normally distributed with mean 0 and variance σ^2; hence the regression relationship meets the specifications for the model.

The data matrix for the regression is given in Table 7.2. Table 7.3 provides the results of the run on the SPSS package [1], including an

Table 7.2 Regression Data Matrix for Table 7.1

Y	X_1	X_2
52	1	0
36	1	0
24	1	0
48	1	0
52	1	0
27	1	0
48	1	0
41	1	0
23	0	1
36	0	1
29	0	1
16	0	1
27	0	1
21	0	0
32	0	0
17	0	0
24	0	0
15	0	0

analysis of variance, the sample regression coefficients, and the variance-covariance matrix of the regression coefficients.

Table 7.3 Results of the Regression Run for the Data of Table 7.2

R SQUARE .51652

F	SIGNIFICANCE
8.01245	.004

ANALYSIS OF VARIANCE	DF	SUM OF SQUARES	MEAN SQUARE
REGRESSION	2	1332.84444	666.42222
RESIDUAL	15	1247.60000	83.17333

VAR	*B*	STD ERROR *B*	*F* / SIGNIFICANCE
X_1	19.200000	5.1991666	13.637507 .002
X_2	4.4000000	5.7679575	.58191728 .457
(CONST)	21.800000	4.0785618	28.569253 .000

VARIANCE/COVARIANCE MATRIX OF THE UNNORMALIZED REGRESSION COEFFICIENTS

b_1	27.03133	
b_2	16.63467	33.26933
	b_1	b_2

The expected sales of design 1 is $\beta_0 + \beta_1$. The point estimate for this parameter is $b_0 + b_1 = 21.8 + 19.2 = 41.0$. Similarly, the expected sales of designs 2 and 3 are estimated by $21.8 + 4.4 = 26.2$, and 21.8, respectively.

Suppose that a 95% confidence interval on the sales of design 1 is desired. By (1.23)

$$S_{b_0+b_1}^2 = S_{b_0}^2 + S_{b_1}^2 + 2S_{b_0 b_1}$$

Unfortunately, $S_{b_0 b_1}$ is not printed out by the SPSS program (although other programs are more cooperative). Actually, $S_{b_0 b_1} = -16.63$. Therefore

$$S^2_{b_0+b_1} = (4.079)^2 + (5.199)^2 + 2(-16.63) = 10.40$$

Hence by (1.13) and using a t distribution for $[b_0 + b_1 - (\beta_0 + \beta_1)]/S_{b_0+b_1}$ with $v = n - m = 18 - 3 = 15$ degrees of freedom:[1]

$$\begin{aligned}\hat{I}_{.95} &= \left[b_0 + b_1 - t_{.025}(S_{b_0+b_1}), b_0 + b_1 + t_{.025}(S_{b_0+b_1})\right] \\ &= \left[41.0 - 2.131\ (3.22), 41.0 + 2.131\ (3.22)\right] \\ &= \left[34.13, 47.87\right]\end{aligned}$$

We could also make more involved estimates. Suppose that we wanted to estimate the difference in expected sales between design 1 and the average of designs 2 and 3. The estimate desired is of

$$\beta_0 + \beta_1 - \frac{(\beta_0 + \beta_2) + \beta_0}{2}$$

or of

$$\beta_1 - \frac{\beta_2}{2}$$

The point estimate is $b_1 - b_2/2 = 19.2 - 2.2 = 17.0$. If an interval is desired, first $S_{b_1 - b_2/2}$ must be found. From (1.23)

$$\begin{aligned}S^2_{b_1-b_2/2} &= S^2_{b_1} + \tfrac{1}{4} S^2_{b_2} - S_{b_1 b_2} \\ &= 27.03 + 33.27/4 - 16.63 \\ &= 18.71\end{aligned}$$

An interval at any desired confidence level can be found by the methods used above; for example,

$$\hat{I}_{.95} = \left[7.78, 26.22\right].$$

As a final part of the analysis suppose that we wish to test the hypothesis that the expected sales for all three designs are identical. In other words, we have

$$\begin{aligned}&H_0 : \beta_0 + \beta_1 = \beta_0 + \beta_2 = \beta_0 \\ &H_1 : \text{above not true}\end{aligned}$$

[1]The use of a t distribution is justified in Section 7.7. However, it should be recalled from Chapter 2 that $(b_j - \beta_j)/S_{b_j}$ has a t distribution with $n - m$ degrees of freedom.

The null hypothesis reduces to the statement that $\beta_1 = \beta_2 = 0$. As shown in Section 2.5, however, this hypothesis is tested by the regression F value. The F for this regression is 8.01, which would be significant if we had chosen the .01 level because the SIGNIFICANCE in Table 7.3 is less than .01. Another way of concluding this would be to find $F_{.01}$ at $\nu_1 = 2$ and $\nu_2 = 15$. Because $F_{.01} = 6.36$, we can reject the null hypothesis. ◀

7.3 THE SINGLE-FACTOR ANOVA MODEL

As shown in this section, what is called the single-factor ANOVA model, despite its somewhat different form and terminology, is merely a linear regression model in disguise. The terminology of ANOVA models reflects the predominant use of these models in the design and analysis of experiments. It should also be noted that the models are sometimes applicable for the analysis of survey data as well.

A *factor* is (usually) a qualitative independent variable. In Example 7.1, there were three factor levels:[1] design 1, design 2, and design 3. The term factor is generally used for an experimental variable, that is, for a variable whose levels are set as part of an experiment.

In models (discussed in Chapter 8) in which there are two or more factors any combination of factor levels is called a *treatment*. In the single-factor model the treatments are equivalent to factor levels. The term single-factor model is somewhat misleading because there are other models (randomized block, latin square) that may have only one factor. The distinction becomes apparent in the next chapter.

Suppose that we are dealing with one factor and that factor has k levels. An experiment is to have r_j observations of the quantitative dependent variable Y for each factor level j. The single-factor ANOVA model states that the ith observation of Y when the jth factor level is operating (or Y_{ij}) is given by

$$Y_{ij} = \mu + \Upsilon_j + \epsilon_{ij} \tag{7.1}$$

where μ is a constant,
Υ_j is called the additive *effect* of the jth factor level on Y,
ϵ_{ij} is the error for observation i on the jth factor level; ϵ is a random variable that has a normal distribution with mean 0 and standard deviation σ.

[1]Note that it was the factor levels that were represented with dummy variables in the regression model. The factor itself was not a regression variable.

It is usual also to impose the restriction

$$\sum_{j=1}^{k} \Upsilon_j = 0 \tag{7.2}$$

This convenient restriction[1] causes the constant μ to have a special interpretation which is explained as follows:

According to the model in (7.1) the expected or "long-term" average value of Y when treatment j is in effect, or μ_{Υ_j}, is

$$\mu_{\Upsilon_j} = E(\mu + \Upsilon_j + \epsilon) = \mu + \Upsilon_j \tag{7.3}$$

The average of the expected values μ_{Υ_j} is equal to

$$\frac{\mu_{\Upsilon_1} + \mu_{\Upsilon_2} + \cdots + \mu_{\Upsilon_k}}{k} = \frac{k\mu + (\Upsilon_1 + \Upsilon_2 + \cdots + \Upsilon_k)}{k}$$

$$= \mu \quad \text{because} \sum_{j=1}^{k} \Upsilon_j = 0$$

As shown in Appendix A, Note 7.1, if restriction (7.2) is absent, unique estimates for μ and Υ_j cannot be found, although $\mu_{\Upsilon_j} = \mu + \Upsilon_j$ can be estimated uniquely. Also, as demonstrated in Note 7.2, other forms of restriction can be used which lead to different interpretations for μ. The following example illustrates this model in terms of the data of Table 7.1.

▶ **Example 7.2** According to (7.1) and (7.2), the observations in Table 7.1 are "caused" by the model parameters and error term as shown in Table 7.4. The factor "package design" has three levels: Υ_1 represents the additive effect due to design 1, Υ_2 is the effect of design 2, and Υ_3 is the effect of design 3.

Table 7.4 The Model's "Explanation" of Table 7.1

Design 1	Design 2	Design 3
$Y_{11} = 52 = \mu + \Upsilon_1 + \epsilon_{11}$	$Y_{12} = 23 = \mu + \Upsilon_2 + \epsilon_{12}$	$Y_{13} = 21 = \mu + \Upsilon_3 + \epsilon_{13}$
$Y_{21} = 36 = \mu + \Upsilon_1 + \epsilon_{21}$	$Y_{22} = 36 = \mu + \Upsilon_2 + \epsilon_{22}$	$Y_{23} = 32 = \mu + \Upsilon_3 + \epsilon_{23}$
$Y_{31} = 24 = \mu + \Upsilon_1 + \epsilon_{31}$	$Y_{32} = 29 = \mu + \Upsilon_2 + \epsilon_{32}$	$Y_{33} = 17 = \mu + \Upsilon_3 + \epsilon_{33}$
$Y_{41} = 48 = \mu + \Upsilon_1 + \epsilon_{41}$	$Y_{42} = 16 = \mu + \Upsilon_2 + \epsilon_{42}$	$Y_{43} = 24 = \mu + \Upsilon_3 + \epsilon_{43}$
$Y_{51} = 52 = \mu + \Upsilon_1 + \epsilon_{51}$	$Y_{52} = 27 = \mu + \Upsilon_2 + \epsilon_{52}$	$Y_{53} = 15 = \mu + \Upsilon_3 + \epsilon_{53}$
$Y_{61} = 27 = \mu + \Upsilon_1 + \epsilon_{61}$		
$Y_{71} = 48 = \mu + \Upsilon_1 + \epsilon_{71}$		
$Y_{81} = 41 = \mu + \Upsilon_1 + \epsilon_{81}$		

[1] An alternative restriction, $\Sigma r_j \Upsilon_j = 0$, is explained in Appendix A, Note 7.2. The note should be read after study of Section 7.6.

It is not difficult to see that if the column under design 1, say, in Table 7.4 were infinitely long its average would be $\mu + \Upsilon_1$ because then the error terms ϵ_{i1} would average out to 0. Hence the expected value for the sales in bags per week of potatoes packaged in design 1 is $\mu + \Upsilon_1$. The constant μ is the average of the expected sales for the three designs. Because the values of ϵ that actually "occurred" can never be known, the parameters μ, Υ_1, Υ_2, and Υ_3 can never be known exactly but must be estimated. ◀

A few words on *randomization* are appropriate at this point. When assigning experimental units to factor levels, we should use some form of randomization device such as a table of random numbers. This is important both for the design of this chapter and for the more complex experimental designs of Chapter 8. We could look on the random assignment of experimental units to factor levels as an attempt to preserve, approximately, the nature of the error term ϵ in the model described by (7.1); for example, if we do not use randomization, the stores assigned to test a particular package design may be particularly receptive to that design (the stores may all be in districts in which mostly younger people live and younger people may like garish designs). Alternatively, a group of stores with particularly diverse characteristics may be assigned to another design. As a result the variance of the error term may be different at different factor levels. If randomization is not followed, deviation from (7.1) is more likely. The consequences of deviation are discussed in the last section of Chapter 8.

The model introduced in this section is known as a *fixed effects model*. This means that the effects Υ_j are parameters and not variables. In contrast, in a *random effects model* the effects are random variables (usually having a normal distribution with mean 0); for example, Table 7.1 could be interpreted as depicting data when the three designs are randomly chosen from an infinitely large family of designs whose expected effect is 0. Such a model would be appropriate to the question "Is package design, in general, an important factor in potato sales?" rather than to a particular interest in design 1, design 2, and design 3.

This book does not deal with random effects models, which often require analysis beyond standard regression methods. Furthermore, fixed effects models are most often used in experimental design in management.

7.4 REGRESSION FORMULATION AND ANOVA OUTPUT

It is possible to obtain the estimates for μ, $\Upsilon_1, \ldots, \Upsilon_k$ simply by representing the factor levels by dummy variables in the regular way and then using a little algebra to extract the estimates. Let the estimators be $\hat{\mu}$, $\hat{\Upsilon}_1, \ldots, \hat{\Upsilon}_k$. The following example illustrates the method:

▶ **Example 7.3** When designs 1 and 2 in Table 7.1, Example 7.1, were represented by dummy variables X_1 and X_2, the regression obtained (Table 7.3) was

$$Y_R = 21.8 + 19.2X_1 + 4.4X_2$$

The expected value of Y when design 1 (or factor level 1) is present is $\beta_0 + \beta_1$ in the regression model; hence

$$\beta_0 + \beta_1 = \mu + \Upsilon_1$$

Similarly,

$$\beta_0 + \beta_2 = \mu + \Upsilon_2$$

$$\beta_0 = \mu + \Upsilon_3$$

By adding the three equations

$$\beta_0 + \beta_1 + \beta_0 + \beta_2 + \beta_0 = \mu + \Upsilon_1 + \mu + \Upsilon_2 + \mu + \Upsilon_3$$

Because

$$\sum \Upsilon_j = 0$$

$$3\beta_0 + \beta_1 + \beta_2 = 3\mu$$

and

$$\mu = \frac{3\beta_0 + \beta_1 + \beta_2}{3}$$

Therefore the estimate $\hat{\mu}$ is

$$\hat{\mu} = \frac{3b_0 + b_1 + b_2}{3}$$

$$= \frac{3(21.8) + 19.2 + 4.4}{3}$$

$$= 29.67$$

Because

$$\Upsilon_1 = \beta_0 + \beta_1 - \mu,$$

$$\hat{\Upsilon}_1 = b_0 + b_1 - \hat{\mu}$$

$$= 21.8 + 19.2 - 29.67$$

$$= 11.33$$

Similarly,

$$\begin{aligned}\hat{\Upsilon}_2 &= b_0 + b_2 - \hat{\mu} \\ &= 21.8 + 4.4 - 29.67 \\ &= -3.47\end{aligned}$$

Also

$$\begin{aligned}\hat{\Upsilon}_3 &= b_0 - \hat{\mu} \\ &= 21.8 - 29.67 \\ &= -7.87\end{aligned}$$

Note that $\hat{\Upsilon}_1 + \hat{\Upsilon}_2 + \hat{\Upsilon}_3 = 0$ (apart from rounding error).

We can interpret $\hat{\mu}$ as the estimated average of the weekly sales capabilities (or populations) of designs 1, 2, and 3. The estimated effect $\hat{\Upsilon}_1$ then means that the average sales of design 1 is estimated to be 11.33 units above this overall average, whereas $\hat{\Upsilon}_2$ and $\hat{\Upsilon}_3$ tell us that the average sales of designs 2 and 3 are estimated to be 3.47 and 7.87 units, respectively, below this overall average. ◀

Although it would be possible to use the algebraic method illustrated above to find ANOVA parameter estimates even for models more complex than the single-factor model, a more convenient method exists. By a rather underhanded use of dummy variables it is possible to obtain $\hat{\mu}$, $\hat{\Upsilon}_1$, $\hat{\Upsilon}_2, \ldots, \hat{\Upsilon}_{k-1}$ directly as coefficients of a regression equation. The last effect $\hat{\Upsilon}_k$ will be obtained from the condition that $\Sigma\hat{\Upsilon}_j = 0$.

The essence of the method is this. A dummy variable is used to represent each of the first $k - 1$ factor levels as before. For observations of the kth factor level, however, the dummy variables, instead of being zero, are all set to -1. As will be seen, this ensures that the regression coefficients estimate the effects Υ_j directly. It also forces β_0 to be equal to μ. This scheme is formalized as follows:

$$Y = \beta_0 + \beta_1 X_1 + \cdots + \beta_{k-1} X_{k-1} + \epsilon \tag{7.4}$$

where $X_j = 1$ if Y is observed under factor level j where $j \neq k$,
$= -1$ if Y is observed under factor level k,
$= 0$ otherwise.

In regression model (7.4)

$$\begin{aligned}\beta_0 &= \mu \\ \beta_1 &= \Upsilon_1 \\ &\vdots \\ \beta_{k-1} &= \Upsilon_{k-1}\end{aligned} \tag{7.5}$$

In order to see why (7.5) must necessarily follow from (7.4), consider a model in which there are four factor levels. The expected value of Y for factor level 1 (or μ_{Υ_1}) is $\mu + \Upsilon_1$ in the ANOVA model. In the regression model (7.4) X_1 will be 1 and $X_2 = X_3 = 0$, hence the expected value is $\beta_0 + \beta_1$. Comparisons that include all the factor levels are summarized in Table 7.5.

Table 7.5 Expected Values of Y for the ANOVA and Regression Models when $k = 4$

	Expected values of Y for ANOVA model (7.1)	Values of dummy variables			Expected values of Y for the regression model (7.4)
		X_1	X_2	X_3	
First factor level	$\mu + \Upsilon_1$	1	0	0	$\beta_0 + \beta_1$
Second factor level	$\mu + \Upsilon_2$	0	1	0	$\beta_0 + \beta_2$
Third factor level	$\mu + \Upsilon_3$	0	0	1	$\beta_0 + \beta_3$
Fourth factor level	$\mu + \Upsilon_4$	−1	−1	−1	$\beta_0 - \beta_1 - \beta_2 - \beta_3$
Total	4μ				$4\beta_0$

Because expected values in the two models must be equal (they are actually the same), it follows that the sum 4μ is equal to the sum $4\beta_0$ and $\mu = \beta_0$. It is then evident that

$$\Upsilon_1 = \beta_1$$

$$\Upsilon_2 = \beta_2$$

$$\Upsilon_3 = \beta_3$$

$$\Upsilon_4 = -\beta_1 - \beta_2 - \beta_3$$

The estimators $\hat{\mu}$, $\hat{\Upsilon}_1$, $\hat{\Upsilon}_2$, $\hat{\Upsilon}_3$, and $\hat{\Upsilon}_4$ are therefore equal to b_0, b_1, b_2, b_3, and $(-b_1 - b_2 - b_3)$, respectively.

▶ **Example 7.4** The experiment in Example 7.1 was analyzed by regression, this time using not the data matrix in Table 7.2 but the data matrix in Table 7.6, which was constructed according to (7.4). Table 7.7 gives the results of the regression run.

The estimated regression relationship is

$$\hat{Y}_R = 29.67 + 11.33X_1 - 3.47X_2$$

Table 7.6 Regression Data Matrix for Table 7.1, According to (7.4)

Y	X_1	X_2
52	1	0
36	1	0
24	1	0
48	1	0
52	1	0
27	1	0
48	1	0
41	1	0
23	0	1
36	0	1
29	0	1
16	0	1
27	0	1
21	− 1	− 1
32	− 1	− 1
17	− 1	− 1
24	− 1	− 1
15	− 1	− 1

Table 7.7 Results of the Regression Run for the Data Matrix on Table 7.6

R SQUARE .51652

F	SIGNIFICANCE
8.01245	.004

ANALYSIS OF VARIANCE	DF	SUM OF SQUARES	MEAN SQUARE
REGRESSION	2	1332.84444	666.42222
RESIDUAL	15	1247.60000	83.17333

VARIABLE	B	STD ERROR B	F / SIGNIFICANCE
X_1	11.333333	2.8839787	15.442984 .001
X_2	− 3.466667	3.2243862	1.1559260 .299
(CONSTANT)	29.666667	2.2026751	181.39972 .000

VARIANCE/COVARIANCE MATRIX OF THE UNNORMALIZED REGRESSION COEFFICIENTS

	b_1	b_2
b_1	8.31733	
b_2	− 4.15867	10.39667

As a result of (7.5),

$$\hat{\mu} = 29.67,$$
$$\hat{\Upsilon}_1 = 11.33,$$
$$\hat{\Upsilon}_2 = -3.47$$

also

$$\hat{\Upsilon}_3 = -11.33 + 3.47 = -7.86.$$

These estimates are the same as those obtained in Example 7.3.

Note that R^2, F, and the analysis of variance printout are identical to those in Table 7.3.

In Example 7.1 a 95% confidence interval on the sales of design 1 (factor level 1) was calculated. The expected sales of design 1 is $\mu_{\Upsilon_1} = \mu + \Upsilon_1$; the point estimate is $b_0 + b_1 = 29.67 + 11.33 = 41.00$, as in Example 7.1. Also

$$S^2_{b_0+b_1} = S^2_{b_0} + S^2_{b_1} + 2S_{b_0 b_1}$$

The covariance estimate $S_{b_0 b_1}$ is not provided by the SPSS program, but the author is more helpful:

$$S_{b_0 b_1} = -1.386$$

Hence

$$S^2_{b_0+b_1} = (2.203)^2 + 8.317 + 2(-1.386)$$
$$= 10.40$$

The interval $\hat{I}_{.95} = [34.14, 47.86]$ is now calculated as in Example 7.1.

The estimate of the difference in expected sales between design 1 and the average of designs 2 and 3 was also calculated in Example 7.1. This difference is

$$\mu_{\Upsilon_1} - \frac{\mu_{\Upsilon_2} + \mu_{\Upsilon_3}}{2}$$

In terms of (7.5)

$$\mu_{\Upsilon_1} - \frac{\mu_{\Upsilon_2} + \mu_{\Upsilon_3}}{2} = \beta_0 + \beta_1 - \frac{\beta_0 + \beta_2 + \beta_0 - \beta_1 - \beta_2}{2}$$
$$= \frac{3\beta_1}{2}$$

The point estimate for this difference is $3b_1/2 = 17.00$.

$$S^2_{3b_1/2} = \frac{9}{4} S^2_{b_1}$$

$$= \frac{9}{4}(8.31733)$$

$$= 18.71$$

$$\therefore S_{3b_1/2} = 4.326$$

Any desired confidence interval could now be calculated; for example,

$$\hat{I}_{.90} = [17.00 - t_{.05}(4.326),\ 17.00 + t_{.05}(4.326)] \quad \text{where } \nu = 18 - 3$$

$$= [17.00 - 1.753(4.326),\ 17.00 + 1.753(4.326)]$$

$$= [9.42,\ 24.58]$$

◀

Analysis of the data in Example 7.1 could, of course, be carried out by an ANOVA program rather than by regression. Such programs use partitioning of squares internally but their output is readily recognizable. There are many canned programs and the amount of information they provide varies. Table 7.8 lists the output of ANVAR1 [2]. Comparison with the analysis of variance for the regression output in Table 7.7 should confirm that the results are identical.

Table 7.8 ANOVA Analysis for the Data in Table 7.1

ANALYSIS OF VARIANCE TABLE

GRAND TOTAL = 568 NO. OBS = 18 MEAN = 31.5556

SOURCES	SUM OF SQUARES	DF	MEAN SQUARES
TREATMENTS	1332.84	2	666.422
ERROR	1247.6	15	83.1734
TOTAL	2580.45	17	

F = 8.01244 ON 2 AND 15 DEGREES OF FREEDOM

PROBABILITY OF F >= 8.01244 WITH 2 AND 15 DF IS 4.29380E-03

In ANOVA printouts the sum of squares due to treatments (factor levels) is sometimes called the *sum of squares among treatment means* because it may be expressed algebraically in that form. Similarly, the sum of squares due to error, or the residual sum of squares, is sometimes called the *sum of squares within treatment groups*.

7.5 AN ALTERNATIVE APPROACH TO ESTIMATION

The estimates $\hat{\mu}$, $\hat{\Upsilon}_1, \ldots, \hat{\Upsilon}_k$ can actually be derived rather easily from the data without using regression. The method is presented in an intuitive and nonrigorous manner in this section. The approach discussed is also applicable to ANOVA models more complex than the single-factor model.

Deriving at least some of the estimates from the data is a useful check on whether the data were entered correctly in the program. This section should deepen the reader's understanding of the nature of the single-factor ANOVA model. It will also serve as a vehicle for introducing some theoretical results that will be useful in planning sample size. Let us return to the data in Example 7.1.

▶ **Example 7.5** Table 7.1 is reproduced for convenience as Table 7.9. Averages have been provided for each column. To estimate μ_{Υ_1}, the expected sales for design 1, it seems reasonable to use the average sales for design 1 in the experiment as $\hat{\mu}_{\Upsilon_1}$. Similarly, the average sales for design 2 in the experiment would be $\hat{\mu}_{\Upsilon_2}$ and the average sales for design 3 would be $\hat{\mu}_{\Upsilon_3}$. Hence

$$\hat{\mu}_{\Upsilon_1} = \hat{\mu} + \hat{\Upsilon}_1 = 41.00$$

$$\hat{\mu}_{\Upsilon_2} = \hat{\mu} + \hat{\Upsilon}_2 = 26.20$$

$$\hat{\mu}_{\Upsilon_3} = \hat{\mu} + \hat{\Upsilon}_3 = 21.80$$

Table 7.9 Sales in Number of Bags Per Week

	Design 1	Design 2	Design 3
	52	23	21
	36	36	32
	24	29	17
	48	16	24
	52	27	15
	27		
	48		
	41		
Average	41.00	26.20	21.80

It also seems reasonable to estimate the overall mean μ of the expected values, where $\mu = (\mu_{\Upsilon_1} + \mu_{\Upsilon_2} + \mu_{\Upsilon_3})/3$ by $(\hat{\mu}_{\Upsilon_1} + \hat{\mu}_{\Upsilon_2} + \hat{\mu}_{\Upsilon_3})/3$. Therefore

$$\hat{\mu} = \frac{41.00 + 26.20 + 21.80}{3}$$

$$= 29.67$$

The same result could have been obtained by adding the three equations above and setting $\Sigma\hat{\Upsilon}_j = 0$. The values $\hat{\Upsilon}_1$, $\hat{\Upsilon}_2$, and $\hat{\Upsilon}_3$ are now

$$\hat{\Upsilon}_1 = 41.00 - 29.67 = 11.33$$

$$\hat{\Upsilon}_2 = 26.20 - 29.67 = -3.47$$

$$\hat{\Upsilon}_3 = 21.80 - 29.67 = -7.87$$

These estimates are the same as those provided by regression in Table 7.7. ◀

In general, the estimate $\hat{\mu}_{\Upsilon_j}$ can be found[1] by taking the average of the observations for factor level j, whereas $\hat{\mu}$ can be found by taking the average of the $\hat{\mu}_{\Upsilon_j}$'s. The estimate $\hat{\Upsilon}_j$ can then be found by a little algebra.

The sums of squares due to factor levels and to error can now be found without recourse to regression.

If $\hat{Y}_{Rj} = \hat{\mu} + \hat{\Upsilon}_j = \hat{\mu}_{\Upsilon_j}$, then the sums of squares can be found as in Section 2.3:

$$\text{SS about regression} = \sum_{j=1}^{k} \sum_{i=1}^{r_j} (Y_{ij} - \hat{\mu}_{\Upsilon_j})^2$$

$$\text{SS due to regression} = \sum_{j=1}^{k} \sum_{i=1}^{r_j} \left(\hat{\mu}_{\Upsilon_j} - \bar{Y}\right)^2 = \sum_{j=1}^{k} r_j\left(\hat{\mu}_{\Upsilon_j} - \bar{Y}\right)^2 \quad (7.6)$$

$$\text{SS total} = \sum_{j=1}^{k} \sum_{i=1}^{r_j} \left(Y_{ij} - \bar{Y}\right)^2$$

The formulas in (7.6) are not computationally efficient and better ones exist. However, they could be used for the analysis of small experiments.

[1]Appendix A, Note 7.1, provides proof and formulas especially for those who are fond of such.

An alternative method of calculating the estimated standard error of $\hat{\mu}_{\Upsilon_j}$ will prove to be useful. The statistic $\hat{\mu}_{\Upsilon_j}$ is calculated from the average of the observations under factor level j. According to (7.1), (7.2), and (7.3),

$$\hat{\mu}_{\Upsilon_j} = \frac{\sum_{i=1}^{r_j} (\mu + \Upsilon_j + \epsilon_{ij})}{r_j}$$

$$= \frac{r_j\mu + r_j\Upsilon_j + \sum_{i=1}^{r_j} \epsilon_{ij}}{r_j}$$

Now

$$\text{VAR}(\hat{\mu}_{\Upsilon_j}) = \frac{1}{r_j^2} \text{VAR}\left(r_j\mu + r_j\Upsilon_j + \sum_{i=1}^{r_j} \epsilon_{ij}\right)$$

Because the terms $r_j\mu$ and $r_j\Upsilon_j$ are constants,

$$\text{VAR}(\hat{\mu}_{\Upsilon_j}) = \frac{1}{r_j^2} \cdot \text{VAR}\left(\sum_{i=1}^{r_j} \epsilon_{ij}\right)$$

Note that the error terms in $\hat{\mu}_{\Upsilon_l}$ are statistically independent of those in $\hat{\mu}_{\Upsilon_j} (j \neq l)$; hence $\hat{\mu}_{\Upsilon_j}$ is statistically independent of $\hat{\mu}_{\Upsilon_l}$. Also,

$$\text{VAR}(\hat{\mu}_{\Upsilon_j}) = \frac{1}{r_j^2} \sum_{i=1}^{r_j} \text{VAR}(\epsilon_{ij})$$

$$= \frac{1}{r_j^2} \cdot r_j \cdot \sigma^2$$

$$\text{VAR}(\hat{\mu}_{\Upsilon_j}) = \frac{\sigma^2}{r_j} \tag{7.7}$$

where σ^2 is the variance of the error term.

In Section 2.4 it was explained that the mean square about regression or the mean square due to error (MSE) was an unbiased estimator for σ^2. Therefore

$$S^2_{\hat{\mu}_{\Upsilon_j}} = \frac{\text{MSE}}{r_j} \tag{7.8}$$

The relationship in (7.8) can be used to do calculations of standard errors for estimators frequently needed in single-factor analysis of variance without employing a variance-covariance matrix of regression coefficients.

▶ **Example 7.6** In Example 7.4 the sample variance of $\hat{\mu}_{\Upsilon_1} - (\hat{\mu}_{\Upsilon_2} + \hat{\mu}_{\Upsilon_3})/2$ was required:

$$S^2_{\hat{\mu}_{\Upsilon_1}} = \frac{\text{MSE}}{r_1} = \frac{83.17}{8} = 10.40$$

$$S^2_{\hat{\mu}_{\Upsilon_2}} = \frac{\text{MSE}}{r_2} = \frac{83.17}{5} = 16.63$$

$$S^2_{\hat{\mu}_{\Upsilon_3}} = \frac{\text{MSE}}{r_3} = \frac{83.17}{5} = 16.63$$

Because $\hat{\mu}_{\Upsilon_1}$, $\hat{\mu}_{\Upsilon_2}$, and $\hat{\mu}_{\Upsilon_3}$ are statistically independent, the estimated variance of $\hat{\mu}_{\Upsilon_1} - (\hat{\mu}_{\Upsilon_2} + \hat{\mu}_{\Upsilon_3})/2$ is

$$S^2_{\hat{\mu}_{\Upsilon_1} - (\hat{\mu}_{\Upsilon_2} + \hat{\mu}_{\Upsilon_3})/2} = S^2_{\hat{\mu}_{\Upsilon_1}} + \tfrac{1}{4} S^2_{\hat{\mu}_{\Upsilon_2}} + \tfrac{1}{4} S^2_{\hat{\mu}_{\Upsilon_3}}$$

according to (1.23). Therefore

$$S^2_{\hat{\mu}_{\Upsilon_1} - (\hat{\mu}_{\Upsilon_2} + \hat{\mu}_{\Upsilon_3})/2} = 10.40 + \frac{16.63}{4} + \frac{16.63}{4}$$

$$= 18.71$$

as before. ◀

It is now opportune to introduce a definition that will be quite useful, especially for the Scheffé method of multiple comparisons introduced in the next section. A *contrast* or *comparison*, L, is a linear function of the expected values μ_{Υ_j}, where

$$L = \sum_{j=1}^{k} c_j \mu_{\Upsilon_j} \tag{7.9}$$

and where $\Sigma_{j=1}^{k} c_j = 0$. The c_j's are known constants.

▶ **Example 7.7** Suppose that we are again interested in the difference in expected sales between design 1 and the average of designs 2 and 3. This difference is $\mu_{\Upsilon_1} - (\mu_{\Upsilon_2} + \mu_{\Upsilon_3})/2$ and can be expressed as the contrast

$$L = \mu_{\Upsilon_1} - \tfrac{1}{2}\,\mu_{\Upsilon_2} - \tfrac{1}{2}\,\mu_{\Upsilon_3}$$

Note that $c_1 = 1$, $c_2 = -\frac{1}{2}$, $c_3 = -\frac{1}{2}$; hence $\Sigma_{j=1}^{3} c_j = 0$ as required. ◀

A contrast L is estimated by $\hat{L}$:

$$\hat{L} = \sum_{j=1}^{k} c_j \hat{\mu}_{\Upsilon_j} \tag{7.10}$$

where $\hat{L}$ is an unbiased estimator for L (i.e., $E(\hat{L}) = L$).

$$\text{VAR}(\hat{L}) = \text{VAR}\left(\sum_{j=1}^{k} c_j \hat{\mu}_{\Upsilon_j} \right)$$

Because the $\hat{\mu}_{\Upsilon_j}$'s are statistically independent,

$$\text{VAR}(\hat{L}) = \sum_{j=1}^{k} \text{VAR}(c_j \hat{\mu}_{\Upsilon_j})$$

$$= \sum_{j=1}^{k} c_j^2 \text{VAR}(\hat{\mu}_{\Upsilon_j})$$

By (7.7)

$$\text{VAR}(\hat{L}) = \sum_{j=1}^{k} c_j^2 \cdot \frac{\sigma^2}{r_j} \tag{7.11}$$

and VAR($\hat{L}$) is estimated by $S_{\hat{L}}^2$ where

$$S_{\hat{L}}^2 = \sum_{j=1}^{k} \frac{c_j^2}{r_j} (\text{MSE})$$

$$S_{\hat{L}}^2 = \text{MSE} \sum_{j=1}^{k} \frac{c_j^2}{r_j} \tag{7.12}$$

It can be shown that $(\hat{L} - L)/S_{\hat{L}}$ has a t distribution with $n - m$ degrees of freedom ($n = \Sigma_{j=1}^{k} r_j$; $m = k - 1 + 1 = k$).

Actually it can also be shown that any linear function $\hat{C}$ of $b_0, b_1, \ldots, b_n$ (for a regression representing the ANOVA models discussed in this book) would give $(\hat{C} - E(C))/S_{\hat{C}}$, a t distribution with $n - m$ degrees of freedom. The function $\hat{C}$ need not necessarily represent a contrast. However, the discussion of multiple comparisons in Section 7.6 requires contrasts. Fortunately many of the estimates of practical interest in ANOVA models can be expressed as contrasts.

▶ **Example 7.8** The calculations in Example 7.6 can now be done by (7.12):

$$\hat{L} = \hat{\mu}_{\Upsilon_1} - \tfrac{1}{2}\,\hat{\mu}_{\Upsilon_2} - \tfrac{1}{2}\,\hat{\mu}_{\Upsilon_3}$$

$$S_{\hat{L}}^2 = \text{MSE} \sum_{j=1}^{k} \frac{c_j^2}{r_j}$$

$$= 83.17\left(1^2\left(\tfrac{1}{8}\right) + \left(-\tfrac{1}{2}\right)^2\left(\tfrac{1}{5}\right) + \left(-\tfrac{1}{2}\right)^2\left(\tfrac{1}{5}\right)\right)$$

$$= 18.71$$ ◀

Although up to this point confidence interval estimation rather than hypothesis testing on contrasts has been discussed, hypothesis testing could be done and is straightforward. In fact, rejection of the null hypothesis $H_0 : L = L_0$, where $H_1 : L \neq L_0$, at the α level of significance, is implied[1] if L_0 is not in the $1 - \alpha$ confidence interval for L; for example, if 0 is not in the 95% confidence interval for L, then we can reject $H_0 : L = 0$ in favor of $H_1 : L \neq 0$ at the .05 level of significance.

7.6 MULTIPLE COMPARISONS

In ANOVA models the main interest is often in the comparison of factor level means. Such comparisons have already been made in preceding sections by confidence intervals, using the t distribution. Each estimate however, must stand alone. In order to expand the preceding statement, let us first consider the following analogous example.

Suppose that there are two events, A and B, such that $\Pr(A) = .95$ and $\Pr(B) = .95$. By elementary probability theory[2] the probability that both A and B will occur is less than or equal to .95. Actually, for the case in which A and B are statistically independent this probability is $.95 \times .95 = .90$.

Suppose that there are two events, A and B, such that $\Pr(A) = .95$ and $\Pr(B) = .95$. By elementary probability theory[2] the probability that both A

$$\hat{I}_{1,.95} : \left[61 \leqslant \mu_{\Upsilon_1} - \mu_{\Upsilon_2} \leqslant 71\right]$$

$$\hat{I}_{2,.95} : \left[22 \leqslant \mu_{\Upsilon_2} - \mu_{\Upsilon_3} \leqslant 28\right]$$

[1]See Appendix A, Note 7.3, for a proof.

[2]By (1.14), $\Pr(A, B) = \Pr(A) \cdot \Pr(B|A)$. Because $\Pr(B|A) \leqslant 1$, $\Pr(A, B) \leqslant \Pr(A)$. Similarly, $\Pr(A, B) \leqslant \Pr(B)$.

To have a confidence level in both $\hat{I}_{1,.95}$ and $\hat{I}_{2,.95}$ *simultaneously* (or *as a family*) is to express a confidence that $\hat{I}_{1,.95}$ *and* $\hat{I}_{2,.95}$ is correct. Analogously to Pr(A and B), this confidence level will generally be less than .95. We can use it to express assurance in statements such as $\mu_{\Upsilon_1} > \mu_{\Upsilon_2}$ *and* $\mu_{\Upsilon_2} > \mu_{\Upsilon_3}$. Simultaneous intervals involving expected factor level values often enable the researcher to devise an approximate ranking of factor levels. *Simultaneous inference* also permits the researcher to make confidence intervals or to test hypotheses that were not arrived at a priori but were suggested by the data. Before exploring this subject further, let us discuss the Scheffé method [3] of obtaining simultaneous confidence intervals.[1]

In the preceding section we defined a contrast, L, by (7.9):

$$L = \sum_{j=1}^{k} c_j \mu_{\Upsilon_j}$$

where $\Sigma c_j = 0$. Also from (7.10)

$$\hat{L} = \sum_{j=1}^{k} c_j \hat{\mu}_{\Upsilon_j}$$

The estimate $\hat{L}$ and its estimated standard error $S_{\hat{L}}$ could be obtained from a regression. A confidence interval for any particular $\hat{L}$ would then be

$$\hat{I}_{1-\alpha} = \left[\hat{L} - t_{\alpha/2} S_{\hat{L}},\ \hat{L} + t_{\alpha/2} S_{\hat{L}}\right] \tag{7.13}$$

Suppose, however, that we were interested in finding family confidence intervals for *all* possible contrasts $\hat{L}$. These contrasts would be created by all possible values c_j that satisfy $\Sigma c_j = 0$. In effect, we would have an infinite number of interval statements of the type

$$\widehat{SI}_{1-\alpha} = \left[\hat{L} - SC_\alpha S_{\hat{L}},\ \hat{L} + SC_\alpha S_{\hat{L}}\right] \tag{7.14}$$

According to a derivation by Scheffé, the value of SC_α that gives the family of $\widehat{SI}_{1-\alpha}$ a $(1 - \alpha) \times 100\%$ confidence level is

$$SC_\alpha = \sqrt{(k - 1)F(\alpha;\ k - 1,\ n - m)} \tag{7.15}$$

[1]Also review Appendix A, Note 2.3.

where $k - 1$ is the number of regression dummy variables (used to represent factor levels $j = 1, \ldots, k$),

$F(\alpha; k-1, n-m)$ is an F value such that $\Pr(F \geqslant F(\alpha; k-1, n-m)) = \alpha$, where F is distributed with $\nu_1 = k - 1$ and $\nu_2 = n - m$,

$n - m$ is the degrees of freedom in the residual sum of squares and $n - m = \sum_{j=1}^{k} r_j - k$.

▶ **Example 7.9** Let us return to the well-worn data of the potato-bag experiment in Example 7.1. The computer run in Example 7.4 is used in analysis. The problem is to obtain 95% confidence level estimates for all the differences among the factor level means μ_{Υ_j} simulatneously. In other works, what is required is the family of confidence intervals for the contrasts

$$L_1 = \mu_{\Upsilon_1} - \mu_{\Upsilon_2}$$

$$L_2 = \mu_{\Upsilon_1} - \mu_{\Upsilon_3}$$

$$L_3 = \mu_{\Upsilon_2} - \mu_{\Upsilon_3}$$

which give the "true" differences in the expected sales for the three potato-bag designs. In order to use (7.14) to obtain a family of intervals for L_1, L_2, and L_3, the estimated standard deviations $S_{\hat{L}_1}$, $S_{\hat{L}_2}$ and $S_{\hat{L}_3}$ must be found. By Example 7.4 or Example 7.6

$$\hat{L}_1 = \hat{\mu}_{\Upsilon_1} - \hat{\mu}_{\Upsilon_2} = 41.0 - 26.2 = 14.8$$

$$\hat{L}_2 = \hat{\mu}_{\Upsilon_1} - \hat{\mu}_{\Upsilon_3} = 41.0 - 21.8 = 19.2$$

$$\hat{L}_3 = \hat{\mu}_{\Upsilon_2} - \hat{\mu}_{\Upsilon_3} = 26.2 - 21.8 = 4.4$$

The sample variance $S^2_{\hat{L}_1}$ can be found from Table 7.7.

$$\begin{aligned}
S^2_{\hat{L}_1} &= S^2_{(b_0+b_1-b_0-b_2)} \\
&= S^2_{b_1-b_2} \\
&= S^2_{b_1} + S^2_{b_2} - 2S_{b_1b_2} \\
&= 8.32 + 10.40 - 2(-4.16) \\
&= 27.03
\end{aligned}$$

$$S_{\hat{L}_1} = 5.20$$

Another way is to use (7.12).

$$S_{\hat{L}_1}^2 = \text{MSE} \sum_{j=1}^{3} \frac{c_j^2}{r_j}$$

$$= 83.17\left(\frac{(1)^2}{8} + \frac{(-1)^2}{5} + \frac{0^2}{5}\right)$$

$$= 27.03$$

$$S_{\hat{L}_1} = 5.20$$

Similarly, $S_{\hat{L}_2} = 5.20$, $S_{\hat{L}_3} = 5.68$.

Calculating SC_α from (7.15), we have

$$\begin{aligned} SC_\alpha &= \sqrt{(k-1)F(\alpha; k-1, n-m)} \\ &= \sqrt{(3-1)F(.05; 3-1, 18-3)} \\ &= \sqrt{2(3.68)} \\ &= 2.71 \end{aligned}$$

Hence the three intervals are

$$\widehat{SI}_{1,.95} = [14.8 - 2.71(5.20), 14.8 + 2.71(5.20)] = [.71, 28.89]$$

$$\widehat{SI}_{2,.95} = [19.2 - 2.71(5.20), 19.2 + 2.71(5.20)] = [5.11, 33.29]$$

$$\widehat{SI}_{3,.95} = [4.4 - 2.71(5.68), 4.4 + 2.71(5.68)] = [-10.99, 19.79]$$

First it should be noted that we could have added $\widehat{SI}_{4,.95}, \ldots, \widehat{SI}_{\infty,.95}$, that is, all the other possible contrasts, to this family of intervals and still have been 95% confident that all of them were correct. We could have, for example, added an interval for the contrast $L_4 = (\mu_{T_1} + \mu_{T_3})/2 - \mu_{T_2}$ to the family. Actually, if we were interested in only these three particular intervals, the Scheffé method would give us intervals that have a higher confidence level than specified.

Note that had we been estimating the intervals separately $t_{.025} = 2.131$ at $\nu = 15$ degrees of freedom would have been employed instead of $SC_\alpha = 2.71$. The resulting intervals would have been narrower. ◀

The Scheffé method of multiple comparison is not the only one.[1] In fact, if one were willing to prespecify a relatively small family of intervals or to restrict the type of contrast to be considered, other methods would be likely to provide narrower intervals at the same confidence level. The Scheffé method, however, comes into its own in what has been termed "exploratory analysis" in Chapter 4.

In Sections 3.4 and 4.5 we discussed the difficulties associated with choosing hypotheses on the basis of the data rather than a priori. Similarly, we are not entitled to use a confidence interval for L calculated from a t value or to choose a hypothesis to test if the interval or test was suggested by the data. The reason for this, as before, is that if we have enough potential hypotheses to choose from chance is likely to provide a spurious significance in one or more of them.

If we calculate intervals by the Scheffé method, we are entitled to use an interval or test on a contrast that seems interesting in the analysis of the data. The reason, of course, is that the Scheffé intervals have a confidence level as a group and any single one of them may be chosen and will have at least the group confidence level.

7.7 ESTIMATING SAMPLE SIZE

There are various approaches to planning sample size in experiments that are based on the assumptions of the ANOVA model. The method discussed here consists of finding the numbers r_j that will give approximately the required interval widths for the preselected contrasts of interest. This method requires a prediction for the MSE.

Suppose that we want the $(1 - \alpha) \times 100\%$ confidence interval for some contrast L_i in an experiment with k factor levels. The approximate width w_i, according to (7.12) and (7.13), of this interval is

$$w_i = 2t_{\alpha/2}\left[\text{MSE}\sum_{j=1}^{k}\frac{c_j^2}{r_j}\right]^{1/2} \tag{7.16}$$

where t has $\sum_{j=1}^{k} r_j - k$ degrees of freedom.

Given required widths for a set of contrasts and an estimated MSE, values of r_j can be tried until all the w_i's from (7.16) are less than or equal to the required widths. The process is simplified if all the r_j's can be assumed to be equal.

[1] A good presentation of the Tukey and the Bonferroni methods is given in Neter and Wasserman [4]. For the Bonferroni method see also Appendix A, Note 2.3.

Similarly, if a set of intervals must be estimated simultaneously at the $(1-\alpha) \times 100\%$ confidence level, the approximate width by (7.12), (7.14), and (7.15) is

$$w_i' = 2[(k-1)F(\alpha; k-1, n-m)]^{1/2}\left[\text{MSE} \sum_{j=1}^{k} \frac{c_j^2}{r_j}\right]^{1/2} \tag{7.17}$$

Trial and error on the r_j's would again be used to match the w_i''s to the required widths.

▶ **Example 7.10** An instructor devised a new aptitude test. Two tests of the same type were already in general use. He was particularly interested in estimating the difference between the average of expected scores on the two current tests and the expected score on his new test. He knew that the standard deviation among students on such tests was about 5. He wished to estimate his contrast with an interval having a width of four points at the 95% confidence level.

Let us assume, for simplicity, that the instructor would test an equal number of students with each test.

By (7.16)

$$w = 2t_{\alpha/2}\left[\text{MSE} \sum_{j=1}^{k} \frac{c_j^2}{r_j}\right]^{1/2}$$

Let $L = (\mu_{\Upsilon_1} + \mu_{\Upsilon_2})/2 - \mu_{\Upsilon_3}$ where μ_{Υ_1} and μ_{Υ_2} are the expected values for the two existing tests and μ_{Υ_3} is the expected value for the new test. Hence

$$c_1 = \tfrac{1}{2} \quad c_2 = \tfrac{1}{2} \quad \text{and} \quad c_3 = -1$$

Also, MSE $= 5^2$. Let $r_1 = r_2 = r_3 = r$. Therefore

$$w = 2t_{\alpha/2}\left\{5^2\left(\frac{1}{r}\right)\left[\left(\frac{1}{2}\right)^2 + \left(\frac{1}{2}\right)^2 + (-1)^2\right]\right\}^{1/2}$$

$$= 12.25t_{\alpha/2}\left(\frac{1}{\sqrt{r}}\right)$$

Let us try $r = 30$. Then, since $\nu = 3(30) - 3$,

$$w = 12.25(1.96)\left(\frac{1}{\sqrt{30}}\right)$$

$$w = 4.43$$

This value is too large but not too far wrong. Because the degrees of freedom are large here and should increase, $t_{\alpha/2}$ is stable and we can solve for r.

$$\sqrt{r} = 12.25(1.96/4)$$

$$r = 36$$

At $r = 36$, $\nu = 3(36) - 3$; hence $t_{.025}$ is still 1.96. The instructor, however, would be wise to choose a larger r than 36 in case MSE turned out to be greater than 25. ◀

▶ **Example 7.11** Suppose that the instructor in Example 7.10 wished to estimate all three possible differences between the expected test scores as well as the contrast specified in that example. The intervals for the differences are to have a width of 4. Suppose further that a family confidence level of 95% is required for the set of intervals.

The contrasts to be estimated are

$$L_1 = \frac{\mu_{\Upsilon_1} + \mu_{\Upsilon_2}}{2} - \mu_{\Upsilon_3}$$

$$L_2 = \mu_{\Upsilon_1} - \mu_{\Upsilon_2}$$

$$L_3 = \mu_{\Upsilon_1} - \mu_{\Upsilon_3}$$

$$L_4 = \mu_{\Upsilon_2} - \mu_{\Upsilon_3}$$

Let us assume again that $r = r_1 = r_2 = r_3$. The interval width for L_1 by (7.17) is

$$w_1' = 2\sqrt{(3-1)F(.05;\, 2,\, 3r-3)}\sqrt{5^2\left(\frac{1}{r}\right)\left[\left(\frac{1}{2}\right)^2 + \left(\frac{1}{2}\right)^2 + (-1)^2\right]}$$

$$= 17.32\sqrt{F(.05,\, 2,\, 3r-3)}\left(\frac{1}{\sqrt{r}}\right)$$

The interval widths for w_2', w_3', and w_4' by (7.17) are

$$w_2' = w_3' = w_4' = 2\sqrt{(3-1)F(.05;\, 2,\, 3r-3)}\sqrt{5^2\left(\frac{1}{r}\right)\left[1^2 + (-1)^2\right]}$$

$$= 20\sqrt{F(.05,\, 2,\, 3r-3)}\left(\frac{1}{\sqrt{r}}\right)$$

Because the last three intervals in the family have the larger width, they will determine r. On the assumption that $\nu = 3r - 3$ will be large (more than 100) Table 3 in Appendix C indicates that $F(.05; 2, 3r - 3)$ will be approximately 3.

Hence, solving

$$4 = 20\sqrt{3} \cdot \frac{1}{\sqrt{r}}$$

$$r = \frac{20^2(3)}{4^2}$$

$$= 75$$

Checking: $F(.05; 2, 3(75) - 3) \approx 3.04$. Re-calculation yields $r = 76$. ◀

EXERCISES

7.1 The same elementary accounting course was taught by four teachers. Two of the teachers, A and B, used interactive time-sharing computer programs to aid instruction. Teachers C and D used the dreary conventional method. Students were assigned at random to the four sections of the course. The grades on the common final examination, calculated by an impartial computer, are listed in Table 7.10.

Table 7.10 Examination Grades for an Accounting Course

	Teacher			
	A	*B*	*C*	*D*
	67	79	74	61
	53	71	79	53
	61	85	75	70
	63	73	71	59
	57	86	63	63
	51	69	81	62
	60	89	69	71
	79	75	80	
	69	80	73	
		76		
		74		
Average	62.22	77.91	73.89	62.71

(a) Fit a linear regression relationship, using ordinary dummy variables. State the assumptions behind the model and how they apply to the analysis of this experiment.

(b) Obtain point estimates for the expected scores of teachers A, B, C, and D from the regression.

(c) Test the null hypothesis that the four teachers produce the same grade results. Use the .01 level of significance.

(d) Estimate the difference between the average expected grade of teachers A and B and the average expected grade of teachers C and D. Use a 95% confidence interval. Use the variance-covariance matrix of the regression coefficients to obtain your answer.

7.2 For the experiment in Exercise 7.1 complete the following:

(a) Construct an ANOVA model for the problem.

(b) Fit a linear regression relationship in which the coefficients give the effects directly.

(c) Test the null hypothesis that the four teachers will produce the same grade results at the .05 level of significance. Use the results in (b).

(d) Repeat Exercise 7.1, part (d) by using the regression in (b).

7.3 Run the data on Exercise 7.1 on a standard ANOVA package. Carefully compare the details of the output with the results in Exercises 7.1 and 7.2.

7.4 Using the data in Table 7.10 Exercise 7.1, find $\hat{\mu}$, $\hat{T}_1$, $\hat{T}_2$, and $\hat{T}_3$ by performing operations on the appropriate averages. Check the estimates with the results in Exercise 7.1 or 7.2.

7.5 Three different operators produced three parts of the same type each. Their times in seconds are listed in Table 7.11.

Table 7.11 Part Completion Times in Seconds

Operator 1	Operator 2	Operator 3
10	7	3
9	5	4
8	6	2

Give a 95% confidence interval for the difference between the average of the times of Operators 1 and 2 and the time of Operator 3. *Note*. A complete *numerical* expression for the confidence interval is required, and you must do the problem without using a regression output.

7.6 For the problem in Exercise 7.1 estimate, *simultaneously*, intervals for all of the pairwise differences between expected grades for the four teachers. Employ the Scheffé method of multiple comparisons at the 95% confidence level. Are the pairwise difference intervals actually at the 95% family confidence level? Try to construct an ordering of teachers at the 95% confidence level. Use a regression output in your calculations.

7.7 Suppose that after a preliminary examination of the grades in Table 7.10 of Exercise 7.1 the teachers decided to compare the mean grade of teachers A and D with the mean grade of teachers B and C.

(a) Find the 95% confidence interval for the difference.
(b) Find the 95% confidence interval, but this time assume that the experiment was designed in the first place to estimate this difference.
(c) Test the null hypotheses that the difference is zero under the assumptions first of (a) and then (b).

7.8 The four teachers in Exercise 7.1 intended to repeat the experiment in the next semester. They wished to estimate all the pairwise differences at the 95% confidence level simultaneously. The interval widths were to be no more than three grade points. Find the required numbers of students in each section.

REFERENCES

1. Norman Nie, Dale H. Bent, and C. Hadlai Hull, *Statistical Package for the Social Sciences* (SPSS). New York: McGraw-Hill, 1970.
2. Jerry L. Mulchy, *ANVAR1*, Hewlett-Packard Time-Shared Basic Program Library.
3. Henry Scheffé, *The Analysis of Variance*. New York: Wiley, 1959.
4. John Neter and William Wasserman, *Applied Linear Statistical Models*, Homewood, Ill.: Irwin, 1974.

chapter 8

Regression analysis of selected experimental designs

8.1 THE RANDOMIZED BLOCK MODEL

In the *randomized block model* all observations of Y are first divided into strata or groups called blocks. Ideally, blocks are chosen so that for any factor level (treatment) there is little variation in the experimental observations (Y values) in a given block. As much as possible of the variation in Y in each factor level is to be attributed to differences in additive (positive or negative) effects on Y produced by the blocks. Each block is assumed to generate the same additive effect, regardless of the factor level.

As an example consider a test to measure the appeal of three possible advertisements for a stereo system. The reactions of potential customers could be quite different among stereo enthusiasts than among those "uneducated" in stereo. Assume that this is true. If the customers are assigned at random to the different advertisements, this difference in stereo sophistication will inflate the error term in the regression model and make it more difficult to identify the effects of the advertisements themselves. Assume that the responses of the educated to the same advertisement are roughly similar and that the responses of the uneducated are also roughly similar. Assume also that the change in the appeal measure caused by stereo sophistication is numerically the same for each advertisement; that is, the sophisticated tend to rate each advertisement 10 points higher than the unsophisticated. If we could remove the effect of stereo sophistication from the error term, more precise results could be obtained.

Essentially, blocking is intended to divide the variation in Y for each factor level into two parts. One part is due to blocking and does not enter the error term. The other part is assumed to be entirely a manifestation of a normally distributed error term.

If the experimental units are individuals, blocking may be done on such characteristics as age, sex, training, and aptitude. The blocks can even be the individuals themselves; for example, if four individuals are given a battery of tests, the individuals may constitute the blocks. Similarly, if the experiment is done on stores, blocking may be done on size, location, promotional expenditure, and so on.

The ANOVA model that incorporates blocks is

$$Y_{ilj} = \mu + \rho_l + \Upsilon_j + \epsilon_{ilj} \tag{8.1}$$

where Y_{ilj} is the ith of r observations for block l and factor level j; there are a blocks and, as before, k factor levels,

μ is a constant;

Υ_j is the additive effect of the jth factor level on Y;

ρ_l is the additive effect of the lth block on Y;

ϵ_{ilj} is the error for the ith observation on the lth block and the jth factor level; ϵ is a random variable that has a normal distribution with mean 0 and variance σ^2.

also $\sum_{j=1}^{k} \Upsilon_j = 0$ and $\sum_{l=1}^{a} \rho_l = 0$;

Equation 8.1 should be compared with (7.1). The main difference is that an additive effect, ρ_l, for each of the blocks is now present. We assume for now that there is an equal number of replications, r, for every combination of block and factor level.

As an illustration of the model, consider Table 8.1, in which the expected values for each block and factor level are given (for this example $a = 3$ and $k = 2$). Note that the expected value of the error term is zero; hence the error term does not appear in the table. Two further definitions will be handy:

$$\mu_{\Upsilon_j} = \mu + \Upsilon_j \tag{8.2}$$

$$\mu_{\rho_l} = \mu + \rho_l \tag{8.3}$$

Table 8.1 Expected Values for Block and Factor Level Combinations with Three Blocks and Two Factor Levels

	Factor level		
Block	1	2	Average
1	$\mu + \rho_1 + \Upsilon_1$	$\mu + \rho_1 + \Upsilon_2$	$\mu + \rho_1$
2	$\mu + \rho_2 + \Upsilon_1$	$\mu + \rho_2 + \Upsilon_2$	$\mu + \rho_2$
3	$\mu + \rho_3 + \Upsilon_1$	$\mu + \rho_3 + \Upsilon_2$	$\mu + \rho_3$
Average	$\mu + \Upsilon_1$	$\mu + \Upsilon_2$	μ

Note that (8.2) is identical to (7.3). Consideration of Table 8.1 reveals that for any a and k the column averages would always be μ_{Υ_j}'s and the row averages would always be μ_{ρ_i}'s because the block and factor level effects, respectively, add to zero. This fact will be useful in the intuitive method of estimation explained in Section 8.3.

Note further that μ is the average of the μ_{Υ_j}'s and also of the μ_{ρ_i}'s. The constant μ is again called the *overall mean*.

Equation 8.1 could be formulated as a regression model simply by using $k - 1$ ordinary dummy variables to represent the factor and $a - 1$ dummy variables to represent blocking. The following formulation, however, which is merely an extension of (7.4), will directly estimate most of the parameters in (8.1) by regression coefficients.

$$Y = \beta_0 + \overbrace{\beta_1 X_1 + \cdots + \beta_{a-1} X_{a-1}}^{a-1 \text{ block effects}} + \overbrace{\beta_a X_a + \cdots + \beta_{a+k-2} X_{a+k-2}}^{k-1 \text{ factor level effects}} + \epsilon \tag{8.4}$$

where *for* $j = 1, \ldots, a - 1$

$$\begin{aligned} X_j &= 1 \quad \text{if } Y \text{ is observed under block } j \\ &= -1 \quad \text{if } Y \text{ is observed under block } a \\ &= 0 \quad \text{otherwise;} \end{aligned}$$

for $j = a, \ldots, a + k - 2$

$$\begin{aligned} X_j &= 1 \quad \text{if } Y \text{ is observed under factor level } j - (a - 1) \\ &= -1 \quad \text{if } Y \text{ is observed under factor level } k \\ &= 0 \quad \text{otherwise.} \end{aligned}$$

The conditions defining the dummy variables in (8.4) seem formidable, but are really not. Of the $a - 1$ dummy variables representing block effects each is equal to 1 for the corresponding block in the first $a - 1$ blocks. All are equal to -1 for the ath block. Similarly, of the $k - 1$ dummy variables representing factor levels each is equal to 1 for the corresponding level in the first $k - 1$ factor levels; all are equal to -1 for the kth factor level.

By the same reasoning used to justify (7.5) it could be shown that

$$\begin{aligned} \beta_0 &= \mu & & \\ \beta_1 &= \rho_1 & \beta_a &= \Upsilon_1 \\ \beta_2 &= \rho_2 & \beta_{a+1} &= \Upsilon_2 \\ &\vdots & &\vdots \\ \beta_{a-1} &= \rho_{a-1} & \beta_{a+k-2} &= \Upsilon_{k-1} \end{aligned} \tag{8.5}$$

Estimation and hypothesis testing could then be done by the usual techniques.

▶ **Example 8.1** A manufacturer of electronic calculators decided to test two competing calculators against his own product with respect to ease of operation. Twelve people, judged to be potential users but without previous experience with calculators, were chosen for the experiment. Each was first rated into one of four categories describing their mathematical "sophistication." In category 1 were people with good mathematical training and recent experience in calculation, whereas category 4 described people with a relatively weak mathematical background and little recent experience in arithmetic. Categories 2 and 3 were in between. Three people fell into each category; to each of them a different calculator was assigned at random. Each was given a set of instructions and one-half hour to practice. A timed test on simple arithmetical operations was then given

Table 8.2 Scores on a Computational Efficiency Test

Mathematical sophistication	Calculator		
	1	2	3 (Own)
1	80	93	75
2	73	84	68
3	79	76	63
4	61	69	68

Table 8.3 Regression Data Matrix for Table 8.2 According to (8.4)

	Mathematical sophistication variables			Calculator variables	
Y	X_1	X_2	X_3	X_4	X_5
80	1	0	0	1	0
93	1	0	0	0	1
75	1	0	0	−1	−1
73	0	1	0	1	0
84	0	1	0	0	1
68	0	1	0	−1	−1
79	0	0	1	1	0
76	0	0	1	0	1
63	0	0	1	−1	−1
61	−1	−1	−1	1	0
69	−1	−1	−1	0	1
68	−1	−1	−1	−1	−1

and a score out of 100 was recorded. The results of the experiment are summarized in Table 8.2.

A regression data matrix given in Table 8.3 was constructed according to (8.4).

Note that in this particular example $r = 1$; in other words, there was only one replication per block-factor level combination.

The results of the regression run with the data in Table 8.3 are given in Table 8.4.

The estimated regression relationship is

$$\hat{Y}_R = 74.08 + 8.58X_1 + .917X_2 - 1.42X_3 - .833X_4 + 6.42X_5$$

By (8.5) $\hat{\mu} = 74.08$

$$\hat{\rho}_1 = 8.58 \qquad \hat{\Upsilon}_1 = -.833$$

$$\hat{\rho}_2 = .917 \qquad \hat{\Upsilon}_2 = 6.42$$

$$\hat{\rho}_3 = -1.42 \qquad \hat{\Upsilon}_3 = .833 - 6.42 = -5.58$$

$$\hat{\rho}_4 = -8.58 - .917 + 1.42 = -8.08$$

◀

Table 8.4 Results of the Regression Run for the Data Matrix in Table 8.3

R SQUARE	.78450
F	SIGNIFICANCE
4.36839	.050

ANALYSIS OF VARIANCE	DF	SUM OF SQUARES	MEAN SQUARE
REGRESSION	5	717.75000	143.55000
RESIDUAL	6	197.16667	32.86111

VAR	*B*	STD ERROR *B*	*F* / SIGNIFICANCE
X_1	8.5833333	2.8662306	8.9678783 .024
X_4	−.83333333	2.3402675	.12679628 .734
X_3	− 1.4166667	2.8662306	.24429417 .639
X_5	6.4166667	2.3402675	7.5177515 .034
X_2	.91666667	2.8662306	.10228233 .760
(CONST)	74.083333	1.6548190	2004.1953 .000

► **Example 8.2** Suppose that in Example 8.1 we wished to test the null hypothesis that the factor level effects are all equal to zero, $H_0 : \Upsilon_1 = \Upsilon_2 = \Upsilon_3 = 0$, or to test the null hypothesis that the block effects are all equal to zero, $H_0 : \rho_1 = \rho_2 = \rho_3 = \rho_4 = 0$. The F value of 4.36839 in Table 8.4 tests neither of these hypotheses but does test

$$H_0 : \Upsilon_1 = \Upsilon_2 = \Upsilon_3 = \rho_1 = \rho_2 = \rho_3 = \rho_4 = 0$$

against H_1: effects are not all equal to zero.

The hypotheses of interest can be tested by using the partial F test in (4.3). To provide data for the test two further regressions were run: one with variables X_1, X_2, X_3 only and one with variables X_4, X_5 only. The analysis of variance for both runs is given in Table 8.5.

Table 8.5 Analysis of Variance Printouts for Regression Runs with the Block Variables Alone and the Factor Level Variables Alone

INDEPENDENT VARIABLES ARE X_1, X_2, X_3

ANALYSIS OF VARIANCE	DF	SUM OF SQUARES	MEAN SQUARE
REGRESSION	3	425.58333	141.86111
RESIDUAL	8	489.33333	61.16667

INDEPENDENT VARIABLES ARE X_4, X_5

ANALYSIS OF VARIANCE	DF	SUM OF SQUARES	MEAN SQUARE
REGRESSION	2	292.16667	146.08333
RESIDUAL	9	622.75000	69.19444

Let us first test the hypothesis system:

$$H_0: \Upsilon_1 = \Upsilon_2 = \Upsilon_3 = 0$$
$$H_1: \text{not all equal}$$

Because $\Upsilon_1 = \beta_4$ and $\Upsilon_2 = \beta_5$, dropping the variables X_4 and X_5 from the regression is equivalent to setting $\Upsilon_1 = \Upsilon_2 = 0$. It is also equivalent to setting $\Upsilon_1 = \Upsilon_2 = \Upsilon_3 = 0$ because $\Upsilon_3 = -\Upsilon_1 - \Upsilon_2$. Hence the partial F statistic for dropping X_4 and X_5 tests the appropriate null hypothesis.

$$F = \frac{[\text{unexplained SS}(X_4 \text{ and } X_5 \text{ missing}) - \text{unexplained SS(all variables in)}]/2}{\text{unexplained SS(all variables in)}/(n - m)}$$

where $n - m = 12 - 6 = 6$, the degrees of freedom for the residual sum of squares in Table 8.4.
Unexplained SS(X_4 and X_5 missing) = 489.33 from Table 8.5. Unexplained SS(all variables in) = 197.17 from Table 8.4; hence

$$F = \frac{(489.33 - 197.17)/2}{197.17/6}$$

$$= \frac{(292.17)/2}{32.86}$$

$$= 4.45$$

With $\nu_1 = 2$, $\nu_2 = 6$, $F_{.01} = 10.92$, and $F_{.05} = 5.14$. Similarly, for testing

$$H_0: \rho_1 = \rho_2 = \rho_3 = \rho_4 = 0$$
$$H_1: \text{not all equal}$$

$$F = \frac{(622.75 - 197.17)/3}{197.17/6}$$

$$= \frac{425.6/3}{32.86}$$

$$= 4.32$$

With $\nu_1 = 3$, $\nu_2 = 6$, $F_{.01} = 9.78$, $F_{.05} = 4.76$

According to this experiment, differences in calculators are not significant at the .05 level and neither are differences in mathematical sophistication. Given the very small sample, the result is not surprising. ◀

The randomized block model may not be the appropriate one. If r is greater than one, it is possible to do a lack of fit analysis by the method in Section 5.5 because each observation in each cell will have the same independent variables. One possible variation in the model is the inclusion of interaction terms; for example, there may be an increase in variable Y not only because of the effect of the lth block and jth factor level but because that particular combination of block and factor level itself has an effect. Young people may do particularly well on a test that requires physical stamina. A model that includes such interactions is the two-factor model discussed later. The model developed there is equally applicable to blocking (when $r > 1$).

8.2 ANOVA OUTPUT AND MISSING OBSERVATIONS

The analysis of randomized block designs may also be done by ANOVA programs that internally utilize partitioning of sums of squares. An illustration follows.

▶ **Example 8.3** The data in Table 8.2 was run on ANVAR3 [1], a time-sharing program. Table 8.6 reproduces the output.

Table 8.6 ANOVA Output for the Data in Table 8.2

SOURCE	SUM OF SQUARES	DF	MEAN SQUARE	F RATIO
ROWS	425.594	3	141.865	4.31732
COLS	292.172	2	146.086	4.44579
RESID	197.156	6	32.8594	
TOTAL	914.922	11		

PROBABILITY OF $F >= 4.31732$ WITH 3 AND 6 DF IS 6.05665E-02

PROBABILITY OF $F >= 4.44579$ WITH 2 AND 6 DF IS .065408

Comparison with Example 8.2 shows that what is labeled here as the sum of squares due to rows (blocks) is the increase in the unexplained sum of squares caused by removal of the block variables.[1] Also, the sum of squares due to columns (factor levels) is the increase in the unexplained sum of squares caused by removing the factor level variables. An increase in the unexplained sum of squares of course causes a corresponding decrease in the regression, or explained, sum of squares. This rationalizes the term "source" in Table 8.6. The F tests are identical.

There is, however, one new implication of the ANOVA output in Table 8.6. If we add the sum of squares due to rows to the sum of squares due to columns, we obtain 717.77, which is the sum[1] due to regression in Table 8.4. ◀

Adding the sum of squares due to blocks to the sum of squares due to factor levels gives us the total sum of squares due to the regression in the preceding example. This will always be true tn experiments in which the number of replications for every block-factor level combination is the same. The reason for this lies in the following result:

[1]Note the slight discrepancies in the sums of squares in Tables 8.5 and 8.6. These are due to rounding under the different methods of calculation.

Result If the independent variables $(X_1, \ldots, X_q)$ are divided into two sets S_1 and S_2 such that every variable in S_1 is uncorrelated (or has zero covariance) with every variable in S_2, then the sum of squares added to the regression by entering S_1 is the same regardless of whether none, any, or all of the variables in S_2 are already in the regression. (8.6)

This result is given without proof (see Draper & Smith,[2], pp. 69–71, for a closely related result) but should follow intuitively from the discussion in Chapter 4. The result has consequences first of all on the interpretation of the sums of squares assigned to blocks and factor levels. It also enables us to use fewer regression runs to analyze variance. Finally, it makes clear the consequences of having unequal numbers of replications in the cells (block-factor level combinations) of the randomized block design.

▶ **Example 8.4** Let us examine the correlation matrix of the variables X_1, X_2, X_3, X_4, and X_5 in Table 8.3. This matrix was printed out by the regression program and is given in Table 8.7.

Table 8.7 Correlation Coefficients for the Data Matrix in Table 8.3

CORRELATION COEFFICIENTS					
X_1	.67484				
X_2	.36442	.50000			
X_3	.26994	.50000	.50000		
X_4	.22208	0.00000	0.00000	0.00000	
X_5	.56106	0.00000	0.00000	0.00000	.50000
	Y	X_1	X_2	X_3	X_4

It is evident from Table 8.7 that none of the variables X_1, X_2, or X_3 is correlated with X_4 or X_5. This fact can readily be verified by inspection of Table 8.3. It will be noted that the mean of each column, $\bar{X}_1, \ldots, \bar{X}_5$, is conveniently zero. By (1.17)

$$S_{X_iX_j} = \frac{1}{ark - 1} \sum_{h=1}^{ark} X_{ih}X_{jh}$$

It is easy to check that

$$\sum_{h=1}^{ark} X_{1h}X_{5h} = 0$$

Therefore X_1 and X_5 are not correlated. It should also be readily seen that if we had, say, three observations in each cell the appropriate correlations would still be zero. However, if there were unequal numbers of observations in each cell, the symmetry would be destroyed and the block variables would become correlated to the factor level variables (unless the cell sizes followed a proportional pattern ([3], pp. 613–614).

It will be recalled that X_1, X_2, and X_3 represent "mathematical sophistication" blocks and that X_4 and X_5 represent "calculator-type" factor levels. Result 8.6 tells us that the sum of squares added by the blocking variables is the same, regardless of the presence or absence of the factor-level variables. This means that we could have used the regression run in Table 8.4 and only one of the runs in Table 8.5 to obtain the complete analysis of variance table, as in Table 8.6.

According to (8.6), when the variables X_1, X_2, and X_3 are in the regression, the SS due to regression, or 425.5833, is the sum attributable to blocking. When all the variables are in the regression (Table 8.4), the sum of squares due to regression is 717.75; hence the sum of squares "due" to the factor levels is 717.75 − 425.5833 = 292.17, as in Table 8.6. Only two regression runs[1] need have been used to duplicate Table 8.6.

The allocation of sums of squares to blocks and to factor levels becomes less clear-cut when the number of replications per cell is not constant. To illustrate this phenomenon let us alter the example. Suppose that in the experiment described in Example 8.1 the test score for the person who tested the company's own calculator at mathematical sophistication level 1 was invalidated (for cheating by doing the calculations mentally). The resulting data are given in Table 8.8.

Table 8.8 Scores on a Computational Efficiency Test: One Score Invalidated

Mathematical	Calculator		
Sophistication	1	2	3
1	80	93	Missing
2	73	84	68
3	79	76	63
4	61	69	68

The regression analysis would proceed along the same lines as in Example 8.1; however, row 3 of the data matrix in Table 8.3 would not be there. Three regression runs were performed and the results are given in Table 8.9.

[1]Some sophisticated programs may have "hypothesis testing options" that accomplish the same results without multiple runs.

Table 8.9 Regression Output for Runs with X_1, X_2, X_3, X_4, X_5, with X_1, X_2, X_3 and with X_4, X_5

CORRELATION COEFFICIENTS

X_1	.73151				
X_2	.36460	.55277			
X_3	.27007	.55277	.50000		
X_4	.25168	.18732	0.00000	0.00000	
X_5	.61661	.18732	0.00000	0.00000	.42105
	Y	X_1	X_2	X_3	X_4

VARIABLES X_1, X_2, X_3, X_4, X_5 IN THE REGRESSION

VAR	B	STD ERROR B
X_1	9.6250000	3.7598482
X_4	− 1.1805556	2.6089160
X_5	6.0694444	2.6089160
X_2	.56944444	3.1540257
X_3	− 1.7638889	3.1540257
(CONST)	74.430556	1.9144210

ANALYSIS OF VARIANCE	DF	SUM OF SQUARES	MEAN SQUARE
REGRESSION	5	725.51389	145.10278
RESIDUAL	5	188.48611	37.69722

VARIABLES X_1, X_2, X_3 IN THE REGRESSION

ANALYSIS OF VARIANCE	DF	SUM OF SQUARES	MEAN SQUARE
REGRESSION	3	512.83333	170.94444
RESIDUAL	7	401.16667	57.30952

VARIABLES X_4, X_5 IN THE REGRESSION

ANALYSIS OF VARIANCE	DF	SUM OF SQUARES	MEAN SQUARE
REGRESSION	2	347.58333	173.79167
RESIDUAL	8	566.41667	70.80208

By (8.5)

$$\hat{\rho}_1 = 9.63$$
$$\hat{\rho}_2 = .569$$
$$\hat{\rho}_3 = -1.76$$
$$\hat{\rho}_4 = -9.63 - .569 + 1.76 = -8.43$$
$$\hat{\Upsilon}_1 = -1.18$$
$$\hat{\Upsilon}_2 = 6.07$$
$$\hat{\Upsilon}_3 = +1.18 - 6.07 = -4.89$$

Suppose that we wished to test the null hypothesis that $\Upsilon_1 = \Upsilon_2 = \Upsilon_3 = 0$, given that the block effects are in the model. This is equivalent to testing $H_0 : \beta_4 = \beta_5 = 0$; $H_1 : \beta_4$ and β_5 are not both equal to zero.

According to the partial F test in (4.3)

$$F = \frac{[\text{unexplained SS}(X_4 \text{ and } X_5 \text{ missing}) - \text{unexplained SS(all variables in)}]/2}{\text{unexplained SS(all variables in)}/(11-6)}$$

Using the data from Table 8.9; we obtain

$$F = \frac{(401.17 - 188.49)/2}{188.49/5}$$
$$= 2.82$$

With $\nu_1 = 2$, $\nu_2 = 5$, $F_{.01} = 13.27$, and $F_{.05} = 5.79$. Similarly, for testing $H_0 : \rho_1 = \rho_2 = \rho_3 = \rho_4 = 0$ (or $\beta_1 = \beta_2 = \beta_3 = 0$); $H_1 : \beta_1, \beta_2$ and β_3 are not all equal to zero,

$$F = \frac{(566.42 - 188.49)/3}{188.49/5}$$
$$= 3.34$$

With $\nu_1 = 3$ and $\nu_2 = 5$, $F_{.01} = 12.06$ and $F_{.05} = 5.41$.

We can summarize the results of these two tests. If we assume the presence of block (mathematical sophistication) levels, we cannot reject the null hypothesis that there is no effect on scores of the use of different calculators (at the .05 level of significance). If we assume the presence of factor levels (calculator effects), we cannot reject the null hypothesis that there is no expected difference in scores among different mathematical sophistication levels. This type of analysis is identical to that carried out when there are no missing observations.

However, we can no longer unambiguously allocate a portion of the sum of squares due to regression to the blocks and another portion to the factor levels. When the factor levels are added to the regression with the blocks absent, they increase the sum of squares due to regression (or decrease the unexplained sum of squares) by 347.58333. When the factor-level variables are added with the blocks already present, they increase the sum of squares due to regression by 725.51389 − 512.83333 = 212.68056. This disparity is predictable by (8.6) because, for example, as shown in Table 8.9, X_1 is correlated with X_4 and X_5. Similarly, it is not possible to speak unambiguously of the sum of squares added by the blocks to the sum of squares due to regression because it depends on whether the factor levels are present or absent. A practical consequence is that we cannot take a short cut to the F tests above by running two regressions only, as described in Example 8.4. ◀

ANOVA programs, which obtain their results by partitioning the total sum of squares, have a little difficulty when the number of replications is unequal. If very few observations are missing, some programs will insert artificial observations (Yates method; see [3]) and then grind away. Such results are approximate and the regression approach is preferable if several observations are missing.[1]

8.3 ESTIMATION IN THE RANDOMIZED BLOCK MODEL

As in the single-factor model, the estimates for the overall mean and for the effects in the randomized block model with constant r can be found easily from the data.

▶ **Example 8.5** The data from the experiment in Example 8.1, given in Table 8.2, are reproduced, for convenience, in Table 8.10 with the addition of row and column averages.

Table 8.10 Scores on a Computational Efficiency Test

Mathematical	Calculator			
Sophistication	1	2	3(Own)	Average
1	80	93	75	82.67
2	73	84	68	75.00
3	79	76	63	72.67
4	61	69	68	66.00
Average	73.25	80.50	68.50	74.08

[1]Sometimes, if entire cells are empty, the block and factor level effects may not be uniquely estimable; hence extreme multicollinearity will occur if the coding described in the preceding section is used. Exercise 8.19 gives an example and suggests alternative coding to analyze a modified model.

It seems reasonable that the expected value $\mu_{\Upsilon_1} = \mu + \Upsilon_1$ may be estimated by the average score on calculator 1. Mathematical sophistication does not enter into the average score because these effects are equally represented in the column and because they sum to zero. (Note that μ_{Υ_1} does not represent an average score for the population of calculator users unless this population has equal numbers in each sophistication category.)

Hence

$$\hat{\mu}_{\Upsilon_1} = 73.25 = \hat{\mu} + \hat{\Upsilon}_1$$

Similarly

$$\hat{\mu}_{\Upsilon_2} = 80.50 = \hat{\mu} + \hat{\Upsilon}_2$$

$$\hat{\mu}_{\Upsilon_3} = 68.50 = \hat{\mu} + \hat{\Upsilon}_3$$

Adding,

$$3\hat{\mu} = 222.25 \quad \text{and} \quad \hat{\mu} = 74.08.$$

Solving these equations, we obtain

$$\hat{\Upsilon}_1 = -.83, \hat{\Upsilon}_2 = 6.42, \quad \text{and} \quad \hat{\Upsilon}_3 = -5.58$$

The expected value $\mu_{\rho_1} = \mu + \rho_1$ may be estimated by the average score for mathematical sophistication category 1. The expected values μ_{ρ_2}, μ_{ρ_3}, and μ_{ρ_4} are estimated by the averages of rows 2, 3, and 4, respectively.

The equations

$$\hat{\mu} + \hat{\rho}_1 = 82.67$$

$$\hat{\mu} + \hat{\rho}_2 = 75.00$$

$$\hat{\mu} + \hat{\rho}_3 = 72.67$$

$$\hat{\mu} + \hat{\rho}_4 = 66.00$$

may be solved for $\hat{\rho}_1, \ldots, \hat{\rho}_4$. All the estimates above check with the regression results in Example 8.1. ◀

This approach can be justified by formulas derived like those for single-factor estimation in Appendix A, Note 7.1. The approach is also applicable when there are missing observations (r is unequal), but a little ingenuity must be used; also, the equations obtained are more difficult to solve.

The estimation of contrasts (as defined in Section 7.5), *a priori*, singly, *post hoc*, or simultaneously, requires an estimate for the standard error of these contrasts. Although such standard errors can always be computed from the estimated variance-covariance matrix of regression coefficients, it is sometimes more convenient to use a formula that uses only the MSE [mean square attributed to error (residual)].

First let us consider the estimated variance of $\hat{\mu}_{\Upsilon_j}$. It can be shown that

$$\mathrm{VAR}(\hat{\mu}_{\Upsilon_j}) = \frac{\sigma^2}{ra} \tag{8.7}$$

where σ^2 is the variance of the error term,
r is the number of observations per block for each factor level,
a is the number of blocks.

The derivation of (8.7) is similar to that of (7.7). Instead of r_j observations under factor level j, there are now ra observations. The block effects do not make a difference because they are constants and not random variables. As before, the statistic $\hat{\mu}_{\Upsilon_j}$ is statistically independent of $\hat{\mu}_{\Upsilon_i}$ where $i \neq j$.

Because the MSE is an unbiased estimator for σ^2, the estimate for the variance $\mathrm{VAR}(\hat{\mu}_{\Upsilon_j})$ is

$$S^2_{\hat{\mu}_{\Upsilon_j}} = \frac{\mathrm{MSE}}{ra} \tag{8.8}$$

$$S^2_{\hat{\mu}_{\beta l}} = \frac{\mathrm{MSE}}{rk} \tag{8.9}$$

where k is the number of factor levels. A contrast among factor level means, μ_{Υ_j}, is defined as

$$L = \sum_{j=1}^{k} c_j \mu_{\Upsilon_j} \tag{8.10}$$

where

$$\sum_{j=1}^{k} c_j = 0.$$

The definition is identical to that in (7.9). The contrast L is estimated by $\hat{L}$:

$$\hat{L} = \sum_{j=1}^{k} c_j \hat{\mu}_{\Upsilon_j} \tag{8.11}$$

The estimated variance of L is

$$S_{\hat{L}}^2 = \text{MSE} \sum_{j=1}^{k} \frac{c_j^2}{ra} = \frac{\text{MSE}}{ra} \sum_{j=1}^{k} c_j^2 \tag{8.12}$$

which can be obtained by substituting ra for r_j in the derivation of (7.12).

As before, the statistic $(\hat{L} - L)/S_{\hat{L}}$ has a t distribution with the same degrees of freedom as the residual sum of squares in the corresponding regression model. There are $a - 1$ coefficients for the blocks, $k - 1$ coefficients for the factor levels, and one constant (β_0). Hence $n - m = ark - (a - 1 + k - 1 + 1) = ark - a - k + 1$, or the number of observations minus the number of parameters in the regression model.

For multiple comparison or *post hoc* intervals among factor levels the Scheffé value SC_α corresponding to (7.15) is needed.

$$SC_\alpha = \sqrt{(k - 1)F(\alpha; k - 1, n - m)} \tag{8.13}$$

where

$k - 1$	is the number of dummy variables used to represent factor levels,
$Pr(F \geqslant F(\alpha; k - 1, n - m)) = \alpha$	where F is distributed with $\nu_1 = k - 1$ and $\nu_2 = n - m$.
$n - m$	is the degrees of freedom in the residual sum of squares of the regression; if the number of replications is constant, $\nu_2 = ark - a - k + 1$.

It is useful to recognize a pattern in the formulas for SC_α that have been given so far and that are given in the experimental designs to follow. The degrees of freedom ν_1 always multiplies F in SC_α. It is the number of regression variables used to represent the means that are being contrasted; for example, with k factor level means, $k - 1$ variables were used to represent factor levels. Hence $\nu_1 = k - 1$. The degrees of freedom ν_2 is always the number of degrees of freedom in the residual sum of squares.

▶ **Example 8.6** The calculator manufacturer in Example 8.1 wished to obtain a 95% confidence interval for the difference between the average expected score on calculators 1 and 2 and the expected score on his own calculator, 3. This estimate was intended before the experiment began.

By using the results in Example 8.1 or Example 8.5 we can obtain the point estimate for the difference $\hat{L}$:

$$\hat{L} = \frac{\hat{\mu}_{\Upsilon_1} + \hat{\mu}_{\Upsilon_2}}{2} - \hat{\mu}_{\Upsilon_3}$$

$$= \frac{73.25 + 80.50}{2} - 68.50$$

$$= 76.88 - 68.50$$

$$= 8.38$$

Before calculating the interval, we need to obtain $S_{\hat{L}}$. This can be done in two ways. Recall that by (8.5)

$$\hat{L} = \frac{b_0 + b_4 + b_0 + b_5}{2} - (b_0 - b_4 - b_5)$$

$$= \frac{b_4}{2} + b_4 + \frac{b_5}{2} + b_5$$

$$= \tfrac{3}{2}(b_4 + b_5)$$

Using the estimated variance-covariance matrix for regression coefficients given in Table 8.11, we have

Table 8.11 Variance-Covariance Matrix of Regression Coefficients for the Data Matrix in Table 8.3

b_1	8.21528				
b_2	− 2.73843	8.21528			
b_3	− 2.73843	− 2.73843	8.21528		
b_4	0.00000	0.00000	0.00000	5.47685	
b_5	0.00000	0.00000	0.00000	− 2.73843	5.47685
	b_1	b_2	b_3	b_4	b_5

$$S_{\hat{L}}^2 = S_{3(b_4+b_5)/2}^2$$

$$S_{\hat{L}}^2 = \tfrac{9}{4} S_{b_4+b_5}^2$$

$$= \tfrac{9}{4}\left(S_{b_4}^2 + S_{b_5}^2 + 2S_{b_4 b_5}\right)$$

$$= \tfrac{9}{4}\left[5.477 + 5.477 + 2(-2.738)\right]$$

$$= 12.32$$

Alternatively, using (8.12), we obtain

$$S_{\hat{L}}^2 = \frac{\text{MSE}}{ra} \sum_{j=1}^{k} c_j^2$$

In Table 8.4 MSE = 32.86;

$$\therefore S_{\hat{L}}^2 = \frac{32.86}{(1)(4)} \left[\left(\tfrac{1}{2}\right)^2 + \left(\tfrac{1}{2}\right)^2 + (-1)^2 \right]$$

$$= 12.32$$

Because $\nu = n - m = 12 - (3 + 2 + 1) = 6$, $t_{\alpha/2} = t_{.025} = 2.447$; hence

$$\hat{I}_{.95} = \left[8.38 \pm \sqrt{12.32}\,(2.447) \right]$$

$$= \left[-.21, 16.97 \right]$$

This also means that for $H_0 : (\mu_{\Upsilon_1} + \mu_{\Upsilon_2})/2 - \mu_{\Upsilon_3} = 0$; H_1 : above not true; H_0 could not be rejected at the .05 level of significance. ◀

▶ **Example 8.7** The calculator manufacturer's electronics engineer (Example 8.1) wished to obtain 95% confidence intervals for the expected differences in scores between the firm's calculators and each of calculators 1 and 2. In view of the fact that scores on the firm's calculator were lower than on competing calculators the engineer hoped to convince management that redesign was necessary.[1]

Because these contrasts were chosen *post hoc*, the Scheffé method of obtaining confidence intervals is appropriate (review Section 7.6). The point estimates for the required contrasts are

$$\hat{L}_I = \hat{\mu}_{\Upsilon_1} - \hat{\mu}_{\Upsilon_3} = 73.25 - 68.50 = 4.75$$

$$\hat{L}_2 = \hat{\mu}_{\Upsilon_2} - \hat{\mu}_{\Upsilon_3} = 80.5 - 68.5 = 12.0$$

By (8.12), and knowing that MSE = 32.86, we get

$$S_{\hat{L}_1}^2 = S_{\hat{L}_2}^2 = \frac{32.84}{(1)(4)} \left[1^2 + (-1)^2 \right] = 16.42$$

These estimated variances could also have been obtained from the variance-covariance matrix of regression coefficients because $\hat{L}_1 = b_0 + b_4 -$

[1]I take no responsibility for the lack of common sense displayed by characters in my examples. The examples merely illustrate techniques of analysis.

$(b_0 - b_4 - b_5) = 2b_4 + b_5$ and, similarly, $\hat{L}_2 = b_4 + 2b_5$. The quantities $S_{\hat{L}_1}$ and $S_{\hat{L}_2}$ could then be calculated by using (1.23).

By (8.13) the Scheffé value SC_α is

$$\begin{aligned} SC_{.05} &= \sqrt{(k-1)F(.05;\, k-1,\, n-m)} \\ &= \sqrt{2F(.05,\, 2,\, 12-6)} \\ &= \sqrt{2(5.14)} \\ &= 3.21 \end{aligned}$$

The 95% confidence intervals for L_1 and L_2 are

$$\begin{aligned} \widehat{SI}_{1,\,.95} &= \left[4.76 \pm 3.21\sqrt{16.42}\,\right] \\ &= \left[-8.25,\ 17.75\right] \end{aligned}$$

and

$$\begin{aligned} \widehat{SI}_{2,\,.95} &= \left[12.0 \pm 3.21\sqrt{16.42}\,\right] \\ &= \left[-1.00,\ 25.00\right] \end{aligned}$$

Note that we can claim that both intervals are true simultaneously at the 95% confidence level. We could also have added as many intervals on contrasts as we wished still the "family" would have been true at the 95% confidence level. ◀

Sample-size estimation done by methods similar to those discussed in Section 7.7 involves no new principles. The aim, as before, is to achieve prespecified confidence interval widths. The width of the intervals will depend on the constants a, k, and r and the anticipated MSE. Presumably k, the number of factor levels, is considered fixed. The number of blocks a may or may not be variable; for example, if the blocks are sex, male or female, it would be difficult to add more sexes to increase the sample size. On the other hand, blocks representing average annual income may be increased in number. We should be aware, however, that the purpose of blocking is to reduce the MSE. Therefore the anticipated MSE should be a function of the number of blocks used. If it is not, then all that the blocks will accomplish is the loss of degrees of freedom (given a constant number of observations per factor level).

▶ **Example 8.8** The manufacturer of calculators in Example 8.1 decided to repeat the experiment that compared ease of operation of the two competing calculators with his own. There was dissatisfaction with the size of sample used before and with various details in the conduct of the experiment. The object of the new experiment was to obtain simultaneous 95% confidence intervals with width 4.0 on the differences in expected scores among the calculators. The same four mathematical sophistication blocks would be used and the MSE was anticipated to be about 30 (on the basis of the previous experiment).

The three estimates of interest are for

$$L_1 = \mu_{\Upsilon_1} - \mu_{\Upsilon_2}$$

$$L_2 = \mu_{\Upsilon_2} - \mu_{\Upsilon_3}$$

$$L_3 = \mu_{\Upsilon_1} - \mu_{\Upsilon_3}$$

Because $a = 4$ and by (8.12) the estimated variance of each $L_i (i = 1, 2, 3)$ is

$$S^2_{\hat{L}_i} = \text{MSE} \sum_{j=1}^{k} \frac{c_j^2}{4r}$$

$$= \frac{30}{4r}\left[1^2 + (-1)\right]^2$$

$$= \frac{15}{r}$$

The Scheffé value or $SC_{.05}$ is, by (8.13),

$$SC_{.05} = \sqrt{(k-1)F(.05;\ k-1,\ n-m)}$$

$$= \sqrt{(3-1)F(.05;\ 3-1,\ 3(4r)-4-3+1)}$$

$$= \sqrt{2F(.05;\ 2,\ 12r-6)}$$

The width of each interval is therefore

$$2\left(\frac{15}{r}\right)^{1/2} \cdot \sqrt{2F(.05;\ 2,\ 12r-6)}$$

which should be equal to 4. However, the value of $\nu_2 = 12r - 6$ is a function of r. As a guess, let us assume that r is 3, hence $\nu_2 = 30$. This gives $F(.05; 2, 30) = 2.32$, which can be used to solve for r:

$$4 = 2\left(\frac{15}{r}\right)^{1/2}[2(3.32)]^{1/2}$$

$$r = \frac{15}{4}(6.64)$$

$$= 24.9$$

$$\approx 25$$

The first guess for r was somewhat low and therefore, recalculating,

$$\nu_2 = (12)(25) - 6 = 294 \quad \text{and} \quad F(.05, 2, 294) \approx 3.03$$

hence

$$4 = 2\left(\frac{15}{r}\right)^{1/2}[2(3.03)]^{1/2}$$

$$r = \frac{15}{4}(6.06)$$

$$\approx 23$$

Repeating the calculations would not change r.

It should be recalled that the use of the Scheffé method implies interest in all possible contrasts. If the firm were truly interested only in the specified differences, then r would probably be somewhat larger than necessary (assuming that the MSE was predicted correctly). ◀

8.4 LATIN SQUARE DESIGNS

In the preceding section the use of blocking to reduce the error term was discussed. We can, however, use more than one set of blocks simultaneously; for example, experimental subjects may be classified by three education levels and three annual income levels. To continue the example, suppose that response to three TV programs is to be tested. If we wished a subject of each possible education and income level to watch each program, we would need $3 \times 3 \times 3$, or 27 subjects. A latin square design would use only nine subjects.

A *latin square design* is an experimental design that uses two blocking variables with the same number of blocks, a, each, to test a factor levels. Each factor level appears in each block only once.

▶ **Example 8.9** A supermarket chain decided to study the effect of the "cosmetic" appearance of potatoes on their sales. Three grades of appearance were to be tested under the chain's own brand. All grades were sold in transparent plastic bags. Grade A potatoes were meticulously washed but were unsorted as to size; Grade B were moderately washed and without extremes in size; Grade C were unwashed but of uniform size. Prices were the same. Three stores were chosen for the experiment and three 2-week time periods were allotted to it. The results in sales of five-pound bags are given in Table 8.12.

Table 8.12 Sales of Bags of Potatoes in a Latin Square Design

	Period 1		Period 2		Period 3	
Stores	Grade	Sales	Grade	Sales	Grade	Sales
1	A	75	B	51	C	43
2	B	31	C	32	A	18
3	C	82	A	61	B	33

Note that each factor level (grade) appears only once for each store block and only once for each period block. Each blocking variable has three blocks and there are also three factor levels. The term latin square derives from the conventional use of Latin letters for the factor levels and the fact that the letters form a square. Of course, other latin squares (arrangements of the letters A, B and C) would have satisfied the definition. This topic is discussed in some comments on design in Section 8.6. ◀

The model used to represent a latin square design is a straightforward extension of the model for the randomized block design

$$Y_{lhj} = \mu + \rho_l + \kappa_h + \Upsilon_j + \epsilon_{lhj} \tag{8.14}$$

where Y_{lhj} is the observation in the lth block of the blocking variable ρ, in the hth block of the blocking variable κ, and under the effect of the jth treatment,
μ is a constant,
ρ_l is the effect of the lth block of the ρ blocking variable,
κ_h is the effect of the hth block of the κ blocking variable,
Υ_j is the effect of the jth factor level
ϵ_{lhj} is a normally distributed error with mean 0 and variance σ^2.

Also

$$\sum_{l=1}^{a} \rho_l = 0, \ \sum_{h=1}^{a} \kappa_h = 0, \ \sum_{j=1}^{a} \Upsilon_j = 0$$

The model described in (8.14) has only one replication per block-block-factor level combination. Including additional replications would mean merely that an extra subscript would be added to Y and ϵ. Because the block and factor level effects add to zero, μ is again called the overall mean.

The regression formulation in (8.14) is simply an extension of (8.4) for an additional blocking variable.

$$Y = \beta_0 + \overbrace{\beta_1 X_1 + \cdots + \beta_{a-1} X_{a-1}}^{a-1 \text{ block effects for } \rho} + \overbrace{\beta_a X_a + \cdots + \beta_{2a-2} X_{2a-2}}^{a-1 \text{ block effects for } \kappa}$$
$$+ \overbrace{\beta_{2a-1} X_{2a-1} + \cdots + \beta_{3a-3} X_{3a-3}}^{a-1 \text{ factor level effects}} + \epsilon \qquad (8.15)$$

where *for* $j = 1, \ldots, a-1$

$X_j = 1$ if Y is observed under block j of blocking variable ρ,
$X_j = -1$ if Y is observed under block a of blocking variable ρ,
$X_j = 0$ otherwise;

for $j = a, \ldots, 2a-2$

$X_j = 1$ if Y is observed under block $j-(a-1)$ of blocking variable κ,
$X_j = -1$ if Y is observed under block a of blocking variable κ,
$X_j = 0$ otherwise;

for $j = 2a-1, \ldots, 3a-3$

$X_j = 1$ if Y is observed under factor level $j-(2a-2)$,
$X_j = -1$ if Y is observed under factor level a,
$X_j = 0$ otherwise.

Whether for a block or a factor, each of the first $a-1$ effects is represented by a dummy variable. When the ath effect is active, the dummy variables become equal to -1. The resulting correspondence between (8.14) and (8.15) is

$$\begin{array}{ccc} & \beta_0 = \mu & \\ \beta_1 = \rho_1 & \beta_a = \kappa_1 & \beta_{2a-1} = \Upsilon_1 \\ \vdots & \vdots & \vdots \\ \beta_{a-1} = \rho_{a-1} & \beta_{2a-2} = \kappa_{a-1} & \beta_{3a-3} = \Upsilon_{a-1} \end{array} \qquad (8.16)$$

One useful property of (8.15) is that any independent variable X_j within a block is uncorrelated with any variable in the other block or in the factor. We can convince ourselves of this by studying examples. The consequence of this property [by result (8.6)] is that sums of squares may be assigned unambiguously to each block and to the factor. In addition, fewer regression runs are needed for a hypothesis test that all the factor effects or, alternatively, all the block effects for a particular blocking variable are zero.

▶ **Example 8.10** Let us obtain the estimates of the block and factor effects in model (8.14) from the data of Table 8.12 by the use of regression formulation (8.15). Let the dummy variables X_1 and X_2 represent stores 1 and 2 when equal to one, respectively, and let X_3 and X_4 represent time periods 1 and 2 when equal to one, respectively. When equal to one, let X_5 represent Grade A and X_6 represent Grade B. The regression data matrix is given in Table 8.13.

Table 8.13 Regression Data Matrix for the Data in Table 8.12

Sales	Stores		Periods		Grades	
Y	X_1	X_2	X_3	X_4	X_5	X_6
75	1	0	1	0	1	0
51	1	0	0	1	0	1
43	1	0	−1	−1	−1	−1
31	0	1	1	0	0	1
32	0	1	0	1	−1	−1
18	0	1	−1	−1	1	0
82	−1	−1	1	0	−1	−1
61	−1	−1	0	1	1	0
33	−1	−1	−1	−1	0	1

It is easy to verify that each of the independent variables in Table 8.13 has a mean of zero. By calculating covariance, using (1.17), it is easy to show that each of the store variables has zero correlation with each of the period variables and with each of the grade variables. This knowledge led to the computer runs summarized in Table 8.14.

From the output in Table 8.14 and by (8.16)

$$\hat{\mu} = b_0 = 47.33$$

$$\hat{\rho}_1 = b_1 = 9.00 \qquad \hat{\kappa}_1 = b_3 = 15.33 \qquad \hat{\Upsilon}_1 = b_5 = 4.00$$

$$\hat{\rho}_2 = b_2 = -20.33 \qquad \hat{\kappa}_2 = b_4 = .667 \qquad \hat{\Upsilon}_2 = b_6 = -9.00$$

$$\hat{\rho}_3 = -9 + 20.33 \qquad \hat{\kappa}_3 = -15.33 - .667 \qquad \hat{\Upsilon}_3 = -4.00 + 9.00$$

$$= 11.33 \qquad = -16.00 \qquad = 5.00$$

Table 8.14 Results of Regression Runs with the Data Matrix in Table 8.13

VARIABLES X_1, X_2, X_3, X_4, X_5, X_6 IN THE REGRESSION
R SQUARE .98810

F	SIGNIFICANCE
27.68159	.035

ANALYSIS OF VARIANCE	DF	SUM OF SQUARES	MEAN SQUARE
REGRESSION	6	3709.33333	618.22222
RESIDUAL	2	44.66667	22.33333

VAR	B	STD ERROR B
X_1	9.0000000	2.2277709
X_3	15.333333	2.2277709
X_5	4.0000000	2.2277709
X_4	.66666667	2.2277709
X_2	− 20.333333	2.2277709
X_6	− 9.0000000	2.2277709
(CONST)	47.333333	1.5752719

VARIANCE/COVARIANCE MATRIX OF THE UNNORMALIZED REGRESSION COEFFICIENT

b_1	4.96296					
b_2	− 2.48148	4.96296				
b_3	0.00000	0.00000	4.96296			
b_4	0.00000	0.00000	− 2.48148	4.96296		
b_5	0.00000	0.00000	0.00000	0.00000	4.96296	
b_6	0.00000	0.00000	0.00000	0.00000	− 2.48148	4.96296
	b_1	b_2	b_3	b_4	b_5	b_6

VARIABLES X_1, X_2, X_3, X_4 IN THE REGRESSION

ANALYSIS OF VARIANCE	DF	SUM OF SQUARES	MEAN SQUARE
REGRESSION	4	3343.33333	835.83333
RESIDUAL	4	410.66667	102.66667

VARIABLES X_1, X_2 IN THE REGRESSION

ANALYSIS OF VARIANCE	DF	SUM OF SQUARES	MEAN SQUARE
REGRESSION	2	1868.66667	934.33333
RESIDUAL	6	1885.33333	314.22222

The ρ variable gives the increments (effects) in sales due to stores 1, 2, and 3, respectively. The κ variable gives the increments due to time periods and the Υ variable gives the increments due to grade of potato.

According to Table 8.14, the increase in the regression sum of squares that occurs because of the addition of the grade (Υ_1 and Υ_2) variables is 3709.33 − 3343.33 = 366.00. This is also the decrease in the unexplained sum of squares when X_5 and X_6 are added. Therefore, by (4.3), testing $H_0 : \beta_5 = \beta_6 = 0$ (or $\Upsilon_1 = \Upsilon_2 = 0$ or $\Upsilon_1 = \Upsilon_2 = \Upsilon_3 = 0$); versus H_1: above not true; can be done by the F value

$$F = \frac{366/2}{44.67/2} = 8.19$$

where $\nu_1 = 2$ and $\nu_2 = 2$. If a test at the .05 level of significance were intended, the null hypothesis could not be rejected because $F_{.05} = 19.00$.

The sum of squares due to time periods is 3343.33 − 1868.67 = 1474.67. Because of (8.6), this is the same sum of squares that would have been obtained by taking the difference in sums of squares due to regression in runs with variables X_1, X_2, X_3, X_4, X_5, and X_6 and with variables X_1, X_2, X_5, and X_6. Similarly, the sum of squares due to stores is 1868.67 − 0 = 1868.67. Appropriate tests for the significance of store effects ($H_0 : \rho_1 = \rho_2 = \rho_3 = 0$) or period effects ($H_0 : \kappa_1 = \kappa_2 = \kappa_3 = 0$) could now be conducted by means of (4.3). ◀

The degrees of freedom arising from the partitioning of squares into components attributed to blocks and factor levels can easily be generalized as in Table 8.15. A test of the hypothesis that all of the effects of a blocking variable (or, alternatively, a factor) are equal to zero is done by the statistic

$$F = \frac{\text{mean square due to the blocking variable (factor)}}{\text{mean square error}} \tag{8.17}$$

Table 8.15 Degrees of Freedom for a Latin Square Design

Sum of Squares due to	df
Row blocks	$a - 1$
Column blocks	$a - 1$
Factor levels	$a - 1$
Error	$a^2 - (a - 1 + a - 1 + a - 1 + 1)$
	$= a^2 - 3a + 2$
	$= (a - 1)(a - 2)$
Total	$a^2 - 1$

which has the degrees of freedom indicated by the numerator ($\nu_1 = a - 1$) and by the denominator ($\nu_2 = a^2 - 3a + 2$). The printout of an ANOVA program usually gives the information in Table 8.15 as well as the appropriate F values for testing for the presence of blocking variables or the factor.

▶ **Example 8.11** The data in Table 8.12, Example 8.9, was entered as input to ANVAR2 [4]; the resulting output is given in Table 8.16.

Table 8.16 ANOVA Output for the Data in Table 8.12

ITEM	SUM OF SQUARES	DF	MEAN SQUARE	*F* RATIO
ROWS	1868.67	2	934.334	41.8383
COLS	1474.67	2	737.334	33.0169
TREATS	366	2	183	8.19451
ERROR	44.6641	2	22.332	

PROBABILITY OF $F\rangle = 41.8383$ WITH 2 AND 2 DF IS 2.33436E-02

PROBABILITY OF $F\rangle = 33.0169$ WITH 2 AND 2 DF IS 2.93971E-02

PROBABILITY OF $F\rangle = 8.19451$ WITH 2 AND 2 DF IS .108761

The reader should check the correspondence of Table 8.16 with the calculations in Example 8.10. ◀

8.5 ESTIMATION IN THE LATIN SQUARE DESIGN

The estimation of effects in latin square designs directly from the original data follows the principles set forth for the single-factor and randomized block models. The rules to follow could, of course, also be justified by formulas as done in Appendix A, Note 7.1.

It is convenient to define

$$\begin{aligned} \mu_{\Upsilon_j} &= \mu + \Upsilon_j \\ \mu_{\rho_l} &= \mu + \rho_l \\ \mu_{\kappa_h} &= \mu + \kappa_h \end{aligned} \tag{8.18}$$

However, we should be careful not to misinterpret the expected values in (8.18); for example, μ_{Υ_j} would be an approximate average of a large population subjected to factor level j only if blocking effects were absent or if they were present in equal numbers for each block, so that the relations $\Sigma \rho_l = 0$ and $\Sigma \kappa_h = 0$ would nullify their effects. Analogous comments apply to μ_{ρ_l} and μ_{κ_h}.

▶ **Example 8.12** Table 8.12 is reproduced as Table 8.17, but with row and column averages added. Store, period, and potato-grade effects are to be estimated.

Table 8.17 Sales in Bags of Potatoes in a Latin Square Design (Table 8.12 reproduced)

	Period 1		Period 2		Period 3		
Store	Grade	Sales	Grade	Sales	Grade	Sales	Average
1	A	75	B	51	C	43	56.33
2	B	31	C	32	A	18	27.00
3	C	82	A	61	B	33	58.67
Average		62.67		48.00		31.33	47.33

To estimate μ_{κ_1} the average of the column of sales for period 1 is taken. This estimate is reasonable because all three stores are represented equally (once); hence their effects cancel out because $\Sigma \rho_l = 0$. Similarly, the effects of potato grades are neutralized. Therefore

$$\hat{\mu} + \hat{\kappa}_1 = 62.67$$

Similarly,

$$\hat{\mu} + \hat{\kappa}_2 = 48.00$$

$$\hat{\mu} + \hat{\kappa}_3 = 31.33$$

Adding,

$$3\hat{\mu} = 62.67 + 48.00 + 31.33$$

$$\hat{\mu} = 47.33$$

and immediately,

$$\hat{\kappa}_1 = 15.33, \ \hat{\kappa}_2 = .67 \text{ and } \hat{\kappa}_3 = -16.00$$

The estimated effects $\hat{\rho}_1$, $\hat{\rho}_2$ and $\hat{\rho}_3$ can be obtained by solving

$$\hat{\mu} + \hat{\rho}_1 = 56.33$$

$$\hat{\mu} + \hat{\rho}_2 = 27.00$$

and

$$\hat{\mu} + \hat{\rho}_3 = 58.67$$

To obtain $\hat{\mu}_{\Upsilon_1}$ we take the average of the three sales observations where grade A was sold. Because of the fiendishly clever design of the latin square, each block is represented once in the sum of these three sales; hence block effects sum to zero. Therefore

$$\hat{\mu} + \hat{\Upsilon}_1 = \frac{75 + 61 + 18}{3}$$

Similarly,

$$\hat{\mu} + \hat{\Upsilon}_2 = \frac{31 + 51 + 33}{3}$$

$$\hat{\mu} + \hat{\Upsilon}_3 = \frac{82 + 32 + 43}{3}$$

Solving

$$\hat{\Upsilon}_1 = 4.00, \hat{\Upsilon}_2 = -9.00, \quad \text{and} \quad \hat{\Upsilon}_3 = 5.00$$ ◀

The estimated standard errors of contrasts can be obtained from the variance-covariance matrix of regression coefficients. However, the following results sometimes simplify calculations:

$$\text{VAR}(\hat{\mu}_{\Upsilon_j}) = \text{VAR}(\hat{\mu}_{\rho_l}) = \text{VAR}(\hat{\mu}_{\kappa_h}) = \frac{\sigma^2}{a} \tag{8.19}$$

The derivation of (8.19) would follow the method of derivation for (7.7). It can be summarized as follows: each of $\hat{\mu}_{\Upsilon_j}$, $\hat{\mu}_{\rho_l}$, and $\hat{\mu}_{\kappa_h}$ is the average of a observations. Each observation is produced by a constant plus an error term (i.e., $Y = \mu + \rho_1 + \kappa_2 + \Upsilon_3 + \epsilon$). The variance of each mean is therefore the variance of the average of a error terms. $\text{VAR}(\Sigma_{i=1}^{a}\epsilon_i/a) = a(\text{VAR}(\epsilon))/a^2$ and (8.19) follows.

As usual, the MSE is used to estimate σ^2 and

$$S^2_{\mu_{\Upsilon_j}} = S^2_{\mu_{\rho_l}} = S^2_{\mu_{\kappa_h}} = \frac{\text{MSE}}{a} \tag{8.20}$$

If there were r replications per cell in the latin square, the denominator in (8.20) would be ra.

A contrast among factor level means, μ_{Υ_j}, is

$$L = \sum_{j=1}^{a} c_j \mu_{\Upsilon_j} \tag{8.21}$$

where

$$\sum_{j=1}^{a} c_j = 0$$

and is estimated by

$$\hat{L} = \sum_{j=1}^{a} c_j \hat{\mu}_{T_j}$$

Contrasts for block-level means could also be used.

The estimated variance of $\hat{L}$ is

$$S_{\hat{L}}^2 = \frac{\text{MSE}}{a} \sum_{j=1}^{a} c_j^2 \tag{8.22}$$

which is derived in a way similar to the derivation for (7.12).

The statistic $(\hat{L} - L)/S_{\hat{L}}$ has a t distribution with the degrees of freedom associated with the residual sum of squares. When simultaneous intervals are required, the Scheffé value is

$$SC_{\alpha} = \sqrt{(a - 1)F(\alpha; a - 1; n - m)} \tag{8.23}$$

where $a - 1$ is the number of variables used to represent factor levels, $n - m$ is equal to $a^2 - 3a + 2$ as in Table 8.15.

It should be clear that the Scheffé values for contrasts among the means of either block would also be given by (8.23).

▶ **Example 8.13** An estimate was required of the expected difference between the average sales effects of Grades A and B and the sales of Grade C for the potato sales experiment in Example 8.9. The 95% confidence interval was also required.

This interval could be calculated *a priori* (as if intended before the experiment began) or *post hoc* (as if suggested by the data). For both assumptions, of course, identical point estimates and estimated standard errors of the contrast are obtained.

$$\hat{L} = \frac{\hat{\mu}_{T_1} + \hat{\mu}_{T_2}}{2} - \hat{\mu}_{T_3}$$

Using the effects estimated in Example 8.9, we obtain

$$\hat{L} = \frac{4.00 - 9.00}{2} - 5 = -7.5$$

By correspondence with the regression output in Table 8.14;

$$\hat{L} = \frac{b_5 + b_6}{2} - (-b_5 - b_6)$$
$$= \tfrac{3}{2}(b_5 + b_6)$$

$$S_{\hat{L}}^2 = S_{3(b_5+b_6)/2}^2$$
$$= \tfrac{9}{4} S_{b_5+b_6}^2$$
$$= \tfrac{9}{4}\left(S_{b_5}^2 + S_{b_6}^2 + 2S_{b_5 b_6}\right)$$
$$= \tfrac{9}{4}[4.96 + 4.96 + 2(-2.48)]$$
$$= 11.17$$

Alternatively,

$$\hat{L} = \frac{\hat{\mu}_{\mathrm{T}_1} + \hat{\mu}_{\mathrm{T}_2}}{2} - \hat{\mu}_{\mathrm{T}_3}$$

By (8.22) and because MSE = 22.33 (Table 8.14), we get

$$S_{\hat{L}}^2 = 22.33 \times \tfrac{1}{3}\left[(\tfrac{1}{2})^2 + (\tfrac{1}{2})^2 + (-1)^2\right]$$
$$= 11.17$$
$$S_{\hat{L}} = 3.34$$

If the a priori interval is calculated, then $t_{\alpha/2} = t_{.025} = 4.303$. Because $\nu = 2$ (Table 8.14);

$$\hat{I}_{.95} = [-7.5 \pm (4.303)3.34]$$
$$= [-21.88, 6.88]$$

The *post hoc* interval is calculated by the use of $SC_{.05}$ instead of $t_{.025}$. Using (8.23) we have

$$SC_{.05} = \sqrt{(3-1)F(.05, 3-1, 2)}$$
$$= \sqrt{2 \times 19.00}$$
$$= 6.16$$

Hence

$$\widehat{SI}_{.95} = [-7.5 \pm (6.16)3.34]$$
$$= [-28.07, 13.07]$$

◀

8.6 SOME DESIGN CONSIDERATIONS FOR LATIN SQUARES

In the analysis of a latin square the model (8.14) was used as a basis. The model has additive effects for both blocking variables and for the factor levels. The error values (ϵ's) are assumed to be statistically independent and to have equal variances. Violation of these assumptions can produce seriously misleading results.

Especially troublesome is the possibility of interactions between blocking variables or between a blocking variable and the factor. The term *interaction*, which is explored further in Section 8.8, means, for example, that there might be a special effect because a particular block and a particular factor level (treatment) occurred together. It could be shown ([5], Section 6.2) that the presence of interactions will cause biased estimates for effects. The estimate of the variance of the error term (MSE) is also biased.

When there is more than one replication per cell, it is possible to use the standard "lack of fit" analysis described in Section 5.5. If we cannot reject the null hypothesis that there is no lack of fit, then we can have some measure of confidence that the model is appropriate. If there is only one replication per cell, the method of Section 5.5 cannot be used. There are methods, however, ([3], p. 780), for testing the appropriateness of the "no interactions" assumption even when there is only one replication per cell.

It is usually suggested that the experimenter choose a latin square at random from the many possible latin squares.[1] This process has some theoretical benefits; for example, it could be shown that the bias due to interactions averages out over all possible squares. This benefit is somewhat theoretical because once a particular square is chosen the experimenter must live with the bias. It could also be shown that under some other types of model deviation the randomization of squares brings us closer in practice to the assumption of the normally distributed error term ([6], Section 5.1).

The basic structure of the latin square design is rather restrictive. The number of blocks in each blocking variable must equal the number of factor levels. Small squares provide few degrees of freedom for the estima-

[1]See Appendix A, Note 8.1, for a brief description of the process.

tion of error (a laughable number by regression rules of thumb). Large squares may be impractical; we may have only so many treatments to be studied.

As has been mentioned, it is possible, but only sometimes, to have more than one replication per cell; for example, in the experiment in Example 8.9 further replications with the same stores and periods are clearly not possible. However, two or more latin squares can often be used in rather inventive and useful ways.

▶ **Example 8.14** Suppose that the supermarket chain in Example 8.9 had run an experiment not on three stores but on six stores in the same three 2-week periods. Stores 4, 5, and 6 were known as "suburban" stores, whereas the first three were chosen from stores in heavily urban areas. The data from the experiment is arranged in Table 8.18.

Table 8.18 Sales of Bags of Potatoes in a Double Changeover Design

Stores	Period 1		Period 2			Period 3		
Urban	Grade	Sales	Grade	Carry-over	Sales	Grade	Carry-over	Sales
1	A	75	B	a	51	C	b	43
2	B	31	C	b	32	A	c	18
3	C	82	A	c	61	B	a	33
Suburban								
4	A	62	C	a	39	B	c	35
5	B	41	A	b	48	C	a	28
6	C	60	B	c	38	A	b	58

The experimental design in Table 8.18 is called a *double changeover design*. It incorporates two latin squares. Not only can the degrees of freedom for the MSE be increased over the single square but other effects may now be studied; for example, the effect of "suburbanity" can now be found (Exercises 8.10 and 8.11).

A particular feature of the double changeover design is that it is suitable for the study of *carry-over* effects. Each store has all three grades of potatoes in the three periods, but sales of Grade B, for example, may be different if the preceding grade were A than if it were C. If one grade is a big seller, people could stock up and depress the sales of the following grade. Positive carry-over is also possible. In the design just given every factor level follows every other factor level the same number of times. For each store that carries a sequence of grades there is another that has the sequence in reverse. In Table 8.18 the carry-over effect of a preceding grade is designated by a lower case letter. A regression formulation for estimating these carry-over effects is the topic of Exercise 8.12. ◀

8.7 ANALYSIS OF COVARIANCE

The essential purpose of blocking variables can often be served in another way. Blocks were used to reduce the variance of the error term by being made to accept some of the blame for the variation in the dependent variable. A continuous variable may sometimes be used for the same purpose; for example, if it is suspected that experimental subjects react differently to an advertisement, depending on their age, we could divide them into three blocks: young, middle-aged, and elderly. Alternatively, we could simply add their ages in years to the regression as an independent variable (review Section 5.4). This independent variable is usually called a *concomitant variable*. The relative merits of blocking versus the use of concomitant variables, which is the *analysis of covariance*, are discussed at the end of this section. Actually, blocks and concomitant variables are often used in the same model.

It is convenient to add the concomitant variable as transformed into a deviation from its own sample mean. If a single concomitant variable, X, is used, the term added to the model is

$$\text{concomitant term} = \gamma(X_i - \bar{X}) \tag{8.24}$$

where i is the number of the observation on Y,
γ is a regression parameter.

▶ **Example 8.15** In the experiment testing learning on different calculators (Example 8.1) it was decided that more precise results could have been

Table 8.19 Scores and IQ's of Subjects in a Computational Efficiency Test

Average IQ = $\bar{X}$ = 114.08

	Calculator								
	1			2			Own		
Mathematical	Score	IQ	IQ dev	Score	IQ	IQ dev	Score	IQ	IQ dev
sophistication	Y	X	$X - \bar{X}$	Y	X	$X - \bar{X}$	Y	X	$X - \bar{X}$
1	80	110	− 4.08	93	117	2.92	75	114	−.08
2	73	109	− 5.08	84	120	5.92	68	111	− 3.08
3	79	126	11.92	76	110	− 4.08	68	121	6.92
4	61	102	− 12.08	69	104	− 10.08	63	125	10.92
Average	73.25		− 2.33	80.50		− 1.33	68.50		3.67

obtained if IQ were used as a concomitant variable. The 12 subjects in the experiment were given IQ tests. The results are summarized in Table 8.19.

The experiment is analyzed by means of the randomized block model in Example 8.1. Factor-level effects due to calculators (Υ_j's) and block effects due to sophistication (ρ_l's) are estimated.

The data matrix for the regression is obtained simply by adding a column $X_6 = X - \overline{X}$ to the matrix in Table 8.3. The parameter γ is estimated by b_6. A summary of computer printouts for the example appears in Table 8.20. Comparison with Table 8.4 should be made.

The variables X_1, X_2, and X_3 represented the first three block (sophistication) effects. Therefore

$$\hat{\rho}_1 = b_1 = 8.83$$

$$\hat{\rho}_2 = b_2 = 1.37$$

$$\hat{\rho}_3 = b_3 = -4.37$$

$$\hat{\rho}_4 = -b_1 - b_2 - b_3 = -5.83$$

Similarly

$$\hat{\Upsilon}_1 = b_4 = .569$$

$$\hat{\Upsilon}_2 = b_5 = 7.22$$

$$\hat{\Upsilon}_3 = -b_4 - b_5 = -7.79$$

Also

$$\hat{\mu} = b_0 = 74.08$$

Treatment means can be defined in the usual way; for example, $\hat{\mu}_{\Upsilon_1} = \hat{\mu} + \hat{\Upsilon}_1$. Note, however, that $\hat{\mu}_{\Upsilon_1}$ is no longer estimated by the average score for calculator 1.

$$\begin{aligned} \hat{\mu}_{\Upsilon_1} &= 74.08 + .569 \\ &= 74.65 \\ &\neq 73.25 \end{aligned}$$

The reason is, of course, that the average score for calculator 1 would estimate $\hat{\mu}_{\Upsilon_1}$ only if the average X for the column were equal to $\overline{X}$. In fact, the column average estimates not μ_{Υ_1} but $\mu + \Upsilon_1 + \gamma \times$ (mean deviation of X for factor level 1).

The average of factor level 1 is $73.25 = \hat{\mu}_{\Upsilon_1} + b_6(-2.33)$ and therefore

$$\begin{aligned} \hat{\mu}_{\Upsilon_1} &= 73.25 - .601(-2.33) \\ &= 74.65 \end{aligned}$$

Table 8.20 Regression Output for Runs with X_1, X_2, X_3, X_4, X_5, X_6, with X_1, X_2, X_3, X_6, and X_4, X_5, X_6

VARIABLES $X_1, X_2, X_3, X_4, X_5, X_6$ IN THE REGRESSION
R SQUARE .96970

F	SIGNIFICANCE
26.66510	.001

ANALYSIS OF VARIANCE	DF	SUM OF SQUARES	MEAN SQUARE
REGRESSION	6	887.19034	147.86506
RESIDUAL	5	27.72633	5.54527

VAR	B	STD ERROR B	F / SIGNIFICANCE
X_1	8.8336884	1.1782901	56.205615
			.001
X_4	.56865530	.99425250	.32711827
			.592
X_6	.60085227	.10869773	30.555859
			.003
X_2	1.3673059	1.1802383	1.3421220
			.299
X_5	7.2178030	.97222202	55.116185
			.001
X_3	− 4.3708570	1.2930322	11.426534
			.020
(CONST)	74.083333	.67978337	11876.814
			0.000

VARIANCE/COVARIANCE MATRIX OF THE UNNORMALIZED REGRESSION COEFFIEICIENT

b_1	1.38837					
b_2	−.45841	1.39296				
b_3	−.48631	−.50567	1.67193			
b_4	.01149	.02068	−.13555	.98854		
b_5	.00656	.01182	−.07746	−.42535	.94522	
b_6	.00492	.00886	−.05809	.02757	.01575	.01182
	b_1	b_2	b_3	b_4	b_5	b_6

VARIABLES X_1, X_2, X_3, X_6 IN THE REGRESSION

ANALYSIS OF VARIANCE	DF	SUM OF SQUARES	MEAN SQUARE
REGRESSION	4	480.22122	120.05530
RESIDUAL	7	434.69545	62.09935

VARIABLES X_4, X_5, X_6 IN THE REGRESSION

ANALYSIS OF VARIANCE	DF	SUM OF SQUARES	MEAN SQUARE
REGRESSION	3	488.48319	162.82773
RESIDUAL	8	426.43348	53.30418

Hence the value $\hat{\mu}_{\Upsilon_1}$ is called the *adjusted mean* for calculator 1. Comparability among mean scores for calculators is enhanced when score differences attributable to differences in IQ are compensated for. Similarly, $\hat{\mu}_{\Upsilon_2}$ and $\hat{\mu}_{\Upsilon_3}$, as obtained from the regression, are called adjusted means.

To test the null hypothesis $H_0 : \Upsilon_1 = \Upsilon_2 = \Upsilon_3 = 0$; against H_1 : not all equal to zero; the procedure to be used is identical to that in Example 8.2. The F test [by (4.3)] is

$$F = \frac{[\text{unexplained SS } (X_4 \text{ and } X_5 \text{ missing}) - \text{unexplained SS (all variables in)}]/2}{[\text{unexplained SS (all variables in)}]/(5)}$$

$$= \frac{(434.70 - 27.73)/2}{27.73/5}$$

$$= 36.69$$

Because $F_{.01} = 13.27$ for $\nu_1 = 2$ and $\nu_2 = 5$, the null hypothesis would have been rejected at the .01 level of significance. Note that this conclusion was not reached in Example 8.2, in which the concomitant variable IQ was not used. ◀

The estimation of contrasts involves no new principles either. Because a new variable is introduced, one degree of freedom is lost in the residual or unexplained sum of squares. As a result, if the t distribution is used, ν must be one less, and if Scheffe's SC_α is used the F distribution involved has ν_2 decreased by one [review expressions (7.15), (8.13) and (8.23)].

▶ **Example 8.16** As a result of the covariance analysis of Table 8.19 the calculator manufacturer decided to obtain a 95% confidence interval for the difference between the average expected score on calculators 1 and 2 and the expected score on his own calculator 3.

As in Example 8.6, the point estimate for the difference is

$$\hat{L} = \frac{\hat{\mu}_{\Upsilon_1} + \hat{\mu}_{\Upsilon_2}}{2} - \hat{\mu}_{\Upsilon_3}$$

$$= \frac{b_0 + b_4 + b_0 + b_5}{2} - (b_0 - b_4 - b_5)$$

$$= (3/2)(b_4 + b_5)$$

$$= 11.68$$

$$S_{\hat{L}}^2 = \tfrac{9}{4}\left(S_{b_4}^2 + S_{b_5}^2 + 2S_{b_4 b_5}\right)$$

Using the variance-covariance matrix in Table 8.20; we have

$$S_{\hat{L}}^2 = \tfrac{9}{4}\left[.989 + .945 + 2(-.425)\right]$$

$$= 2.44$$

$$S_{\hat{L}} = 1.56$$

By (8.13)

$$SC_{.05} = \sqrt{(k-1)F(.05;\, k-1,\, n-m)}$$

$$= \sqrt{2F(.05;\, 2,\, 5)}$$

$$= \sqrt{2(5.79)}$$

$$= 3.40$$

$$\widehat{SI}_{.95} = \left[11.68 \pm 1.56(3.40)\right]$$

$$= \left[6.37,\, 16.99\right]$$

Other intervals may be added to the family while maintaining the 95% confidence level. We can also conclude (at the .05 level of significance) that this difference L is not equal to zero. ◀

Covariance analysis is often an efficient method of reducing the experimental error. It is often not possible to choose blocks for every variable that may influence Y. If the experimental subjects are people, certain information about them may become available after they are chosen or even during the experiment. Preexperimental attitude scores are an example.

Also, some variables are quantitative and the nature of their effect on the dependent variable may be known; for example, consider an industrial experiment in which the effects of different chemical treatments on breaking strength of a fiber are studied. Strength is proportional to the fiber's cross-sectional area. The area may be measurable but cannot be absolutely controlled. We could then use a cross-sectional area as a concomitant variable. Note that a concomitant variable may act linearly or nonlinearly but that its form must be specified. As shown in Section 5.4, with blocking variables no such specification of the response of Y need be made. However, blocking variables gobble up more degrees of freedom if there are more than two blocks in the variable.

We must be careful in covariance analysis if the factor levels have a causative effect on the concomitant variable; for example, suppose that three different training methods were used on salesmen as an experiment. After the training, but before the salesmen were released on the customers, someone remembered that self-confidence is important in sales. The trainees were then recalled for a quick written test designed to reveal that trait. The scores were used as a concomitant variable. The results of the experiment seemed to reveal that the type of training was not important (significant) but that self-confidence was. The truth could have been that a by-product of the training programs was varying degrees of self-confidence.

The slope of the concomitant variable must be the same for all the factor levels; otherwise it would not be meaningful to compare factor-level means. This assumption may be tested by allowing different slopes for different factor levels by adding extra terms to the regression, as in Section 5.4. An F test (4.3) could then be used to test the assumption of equal slopes.

8.8 TWO-FACTOR ANOVA MODEL

Experiments in which two or more factors are tested simultaneously are known as *factorial experiments*. This section is devoted to the analysis of two-factor studies. The methods used could be extended to designs with three or more factors.

If there are two factors, they could, of course, be studied one at a time. It can be shown that a two-factor study can accomplish the same precision in estimation with far fewer experimental units. In addition, the two-factor study enables us to estimate interaction effects and to evaluate their importance.

Let us consider a model:

$$Y_{ilj} = \mu + \Phi_l + \Upsilon_j + (\Phi\Upsilon)_{lj} + \epsilon_{ilj} \tag{8.25}$$

where Y_{ilj} is the ith of r observations for factor level l of factor 1 and for level j of factor 2; factor 1 has f levels and factor 2 has k levels; as explained subsequently, $r \geqslant 2$;

μ is a constant;

Φ_l is the additive effect of factor level l for factor 1;

Υ_j is the additive effect of factor level j for factor 2;

$(\Phi\Upsilon)_{lj}$ is the additive effect due to the combination of the lth level of factor 1 and the jth level of factor 2;

ϵ_{ilj} is the error for the ith observation for treatment (l, j);

ϵ has a normal distribution with variance σ^2;

also

$$\sum_{l=1}^{f} \Phi_l = 0, \quad \sum_{j=1}^{k} \Upsilon_j = 0, \quad \sum_{l=1}^{f} (\Phi\Upsilon)_{lj} = 0, \quad \sum_{j=1}^{k} (\Phi\Upsilon)_{lj} = 0$$

and *treatment* (l, j) is the combination of the lth level of factor 1 and the jth level of factor 2.

Model 8.25 follows the same pattern as all of the previously discussed ANOVA models. The factors produce additive effects (positive or negative) that sum to zero. However, there are additional effects over and above factor effects, which are due to particular combinations of factor levels. These *interaction effects* also sum to zero over each factor; without a constraint of this type they could not have been determined unambiguously. The constant μ could again be called the overall mean for reasons that should become obvious. The ANOVA model and its transmutation into a regression model are illustrated in the following example.

▶ **Example 8.17** A manufacturer of a special-purpose glue wished to test the effectiveness of three alternative chemical compound additives in producing adhesive strength (measured in kilograms per square centimeter). At the same time three alternative curing temperatures in the manufacturing process were tested. It was suspected that the relative effectiveness of the compounds might be different at different temperatures. Twenty-seven batches were prepared, three assigned to each of the nine possible treatments. Table 8.21 shows the adhesive strength (average) of each batch.

Table 8.21 Effects of Compounds and Curing Temperatures on Adhesive Strength

	Compound		
Temperature	1	2	3
1	138	154	81
	130	139	70
	135	160	75
2	81	111	49
	93	100	39
	79	104	55
3	95	141	51
	100	137	63
	93	128	72

Let us use the effects in (8.25) to "explain" the observations in Table 8.21; the error term is ignored for the time being. Let the additive effects due to temperature be Φ_1, Φ_2, and Φ_3. Note that $\Phi_1 + \Phi_2 + \Phi_3 = 0$. Let the effects due to the chemical compounds be Υ_1, Υ_2, and Υ_3, which also sum to zero. Each temperature effect Φ_l is the same regardless of the chemical compound, and each compound has the same effect Υ_j, regardless of the temperature. The interaction effects $(\Phi\Upsilon)_{lj}$ are potentially different for each treatment (l, j). Table 8.22 summarizes these effects.

Table 8.22 Summary of the Effects Assumed by Model 8.25 to be Acting in Table 8.21

Effects of temperature (for each compound)	Effects of compounds (for each temperature)		
	Υ_1	Υ_2	$\Upsilon_3 = -\Upsilon_1 - \Upsilon_2$
Φ_1	$(\Phi\Upsilon)_{11}$	$(\Phi\Upsilon)_{12}$	$(\Phi\Upsilon)_{13} = -(\Phi\Upsilon)_{11} - (\Phi\Upsilon)_{12}$
Φ_2	$(\Phi\Upsilon)_{21}$	$(\Phi\Upsilon)_{22}$	$(\Phi\Upsilon)_{23} = -(\Phi\Upsilon)_{21} - (\Phi\Upsilon)_{22}$
$\Phi_3 = -\Phi_1 - \Phi_2$	$(\Phi\Upsilon)_{31} = -(\Phi\Upsilon)_{11} - (\Phi\Upsilon)_{21}$	$(\Phi\Upsilon)_{32} = -(\Phi\Upsilon)_{12} - (\Phi\Upsilon)_{22}$	$(\Phi\Upsilon)_{33} = +(\Phi\Upsilon)_{11} + (\Phi\Upsilon)_{12} + (\Phi\Upsilon)_{21} + (\Phi\Upsilon)_{22}$

Note that because of the constraints applied to (8.25) the effects in the last column and the last row of Table 8.22 can be expressed in terms of the effects in the other rows and columns; for example, because $(\Phi\Upsilon)_{11} + (\Phi\Upsilon)_{12} + (\Phi\Upsilon)_{13} = 0$, it follows that $(\Phi\Upsilon)_{13} = -(\Phi\Upsilon)_{11} - (\Phi\Upsilon)_{12}$. If we wish the coefficients of dummy variables in a regression formulation to represent the effects directly, dummy variables can be coded as in Table 8.23.

Table 8.23 Dummy Variable Coding Suggested by Table 8.22 (Unless specified, $X_j = 0$)

	Compound		
Temperatures	1 $X_3 = 1$	2 $X_4 = 1$	3 $X_3 = -1, X_4 = -1$
1 $X_1 = 1$	$X_5 = 1$	$X_6 = 1$	$X_5 = -1$ $X_6 = -1$
2 $X_2 = 1$	$X_7 = 1$	$X_8 = 1$	$X_7 = -1$ $X_8 = -1$
3 $X_1 = -1$ $X_2 = -1$	$X_5 = -1$ $X_7 = -1$	$X_6 = -1$ $X_8 = -1$	$X_5 = 1$ $X_6 = 1$ $X_7 = 1$ $X_8 = 1$

Coding for the effects Φ_l and Υ_j is, of course, identical to that used in the randomized block model (8.4). In fact, the coding for a two-factor model with no interactions is the same as that for a randomized block model. Factor levels are coded just like blocks.

The dummy variables for the interaction effects are suggested by the interactions represented within the cells of Table 8.22; for example, the coefficient of X_5 will be $(\Phi\Upsilon)_{11}$ and the coefficient of X_6 will be $(\Phi\Upsilon)_{12}$. Therefore, when X_5 and X_6 are both -1, the effect within the model will be $-(\Phi\Upsilon)_{11} - (\Phi\Upsilon)_{12}$, as required.

Table 8.24 gives the resulting regression data matrix:

Table 8.24 Regression Data Matrix for Table 8.21 Using the Coding of Table 8.23

	Temperature variables		Chemical compound variables		Interaction variables			
Y	X_1	X_2	X_3	X_4	X_5	X_6	X_7	X_8
138	1	0	1	0	1	0	0	0
130	1	0	1	0	1	0	0	0
135	1	0	1	0	1	0	0	0
154	1	0	0	1	0	1	0	0
139	1	0	0	1	0	1	0	0
160	1	0	0	1	0	1	0	0
81	1	0	−1	−1	−1	−1	0	0
70	1	0	−1	−1	−1	−1	0	0
75	1	0	−1	−1	−1	−1	0	0
81	0	1	1	0	0	0	1	0
93	0	1	1	0	0	0	1	0
79	0	1	1	0	0	0	1	0
111	0	1	0	1	0	0	0	1
100	0	1	0	1	0	0	0	1
104	0	1	0	1	0	0	0	1
49	0	1	−1	−1	0	0	−1	−1
39	0	1	−1	−1	0	0	−1	−1
55	0	1	−1	−1	0	0	−1	−1
95	−1	−1	1	0	−1	0	−1	0
100	−1	−1	1	0	−1	0	−1	0
93	−1	−1	1	0	−1	0	−1	0
141	−1	−1	0	1	0	−1	0	−1
137	−1	−1	0	1	0	−1	0	−1
128	−1	−1	0	1	0	−1	0	−1
51	−1	−1	−1	−1	1	1	1	1
63	−1	−1	−1	−1	1	1	1	1
72	−1	−1	−1	−1	1	1	1	1

The results of regression runs with the data matrix of Table 8.24 are given in Table 8.25. At this point, however, two important observations are made. First, it is easy to verify that

$$X_5 = X_1X_3$$

$$X_6 = X_1X_4$$

$$X_7 = X_2X_3$$

and

$$X_8 = X_2X_4$$

The pattern behind these relationships should be evident from Table 8.23. This result generalized below (in (8.26)) provides a quick way of obtaining the coding for the interactions.

The other observation is that every variable representing temperature is uncorrelated with every variable representing chemical compounds and with every variable representing interactions. This, of course, would be evident from a computer printout of the correlation matrix, but the claim is verifiable by inspection because each independent variable in Table 8.24 conveniently sums to zero. Therefore, if sums such as ΣX_iX_j are equal to zero, then X_i and X_j have a correlation coefficient of zero. This result is also generalized (in (8.27)). ◀

Result If $X_1, \ldots, X_{f-1}$ are used to represent the first $f - 1$ factor levels of factor 1 and $X_f, \ldots, X_{f+k-2}$ are used to represent the first $k - 1$ factor levels of factor 2, the interaction in treatment (i, j) is represented by X_iX_{f-1+j} for $i = 1, \ldots, f - 1$; $j = 1, \ldots, k - 1$. (8.26)

Result 8.26 is a useful shortcut. However, the approach illustrated in Table 8.22 allows more flexibility in specifying what effects are to appear directly as regression coefficients (see Exercise 8.16).

Result If the number of replications per treatment is constant, every variable representing a factor level will have zero correlation[1] with every variable representing levels in the other factor and with every variable representing interactions. (8.27)

[1] Result 8.27 holds for the type of coding used in this chapter, but if we used "ordinary" dummy variable coding, the result would not hold: For example, we could use zeros wherever − 1's appear in Table 8.24. This corresponds to the way dummy variables were employed in Example 7.1. The correlation between each of X_5, X_6, X_7, and X_8 and each of X_1, X_2, X_3, and X_4 would be nonzero. With such coding estimation of effects and hypothesis testing is somewhat more difficult ([3], Section 19.4).

The significance of Result 8.27 in conjunction with Result 8.6 is that the regression sums of squares can be unambiguously allocated among the factors and the interactions.

▶ **Example 8.18** The data on Table 8.24, Example 8.17, was run on IDA [7], an interactive program that performs regression analysis. The results are summarized in Table 8.25.

Table 8.25 Regression Output for Runs with $X_1, X_2, X_3, X_4, X_5, X_6, X_7, X_8$, with X_1, X_2 and with X_3, X_4

VARIABLES $X_1, X_2, X_3, X_4, X_5, X_6, X_7, X_8$ IN THE REGRESSION

	MULTIPLE R	R-SQUARE
UNADJUSTED	0.9843	0.9689
ADJUSTED	0.9773	0.9551

STD ERROR OF RESIDUALS = 7.34587

SOURCE	SUM OF SQUARES	DF	MEAN SQUARE	F RATIO
REGRESSION	3.02907E+04	8	3.78634E+03	70.17
RESIDUALS	9.71313E+02	18	5.39618E+01	
TOTAL	3.12620E+04	26	1.20238E+03	

VAR	B(STD V)	B	STD ERROR(B)	T RATIO
X_1	0.5092	2.1222E+01	1.9993E+00	10.615
X_2	– 0.4799	–.2000E+02	1.9993E+00	– 10.004
X_3	0.1413	5.8889E+00	1.9993E+00	2.945
X_4	0.7545	3.1444E+01	1.9993E+00	15.728
X_5	0.1611	8.2222E+00	2.8274E+00	2.908
X_6	– 0.0131	–.6667E+00	2.8274E+00	– 0.236
X_7	– 0.0109	–.5556E+00	2.8274E+00	– 0.196
X_8	– 0.1067	–.5444E+01	2.8274E+00	– 1.926
CONST		9.9000E+01	1.4137E+00	70.028

VARIABLES X_1, X_2 IN THE REGRESSION

SOURCE	SUM OF SQUARES	DF	MEAN SQUARE	F RATIO
REGRESSION	7.66689E+03	2	3.83345E+03	3.90
RESIDUALS	2.35951E+04	24	9.83129E+02	
TOTAL	3.12620E+04	26	1.20238E+03	

Table 8.25 (Continued)

VARIABLES X_3, X_4 IN THE REGRESSION

SOURCE	SUM OF SQUARES	DF	MEAN SQUARE	*F* RATIO
REGRESSION	2.17549E+04	2	1.08775E+04	27.46
RESIDUALS	9.50709E+03	24	3.96129E+02	
TOTAL	3.12620E+04	26	1.20238E+03	

From Table 8.25

$$\hat{\mu} = b_0 = 99.00$$

$$\hat{\Phi}_1 = b_1 = 21.22$$

$$\hat{\Phi}_2 = b_2 = -20.00$$

$$\hat{\Phi}_3 = -21.22 + 20.00 = -1.22$$

$$\hat{\Upsilon}_1 = b_3 = 5.89$$

$$\hat{\Upsilon}_2 = b_4 = 31.44$$

$$\hat{\Upsilon}_3 = -5.89 - 31.44 = -37.33$$

$$\widehat{(\Phi\Upsilon)}_{11} = b_5 = 8.22$$

$$\widehat{(\Phi\Upsilon)}_{12} = b_6 = -.67$$

$$\widehat{(\Phi\Upsilon)}_{13} = -8.22 + .67 = -7.55$$

$$\widehat{(\Phi\Upsilon)}_{21} = b_7 = -.56$$

$$\widehat{(\Phi\Upsilon)}_{22} = b_8 = -5.44$$

$$\widehat{(\Phi\Upsilon)}_{23} = +.56 + 5.44 = 6.00$$

$$\widehat{(\Phi\Upsilon)}_{31} = -8.22 + .56 = -7.66$$

$$\widehat{(\Phi\Upsilon)}_{32} = +.67 + 5.44 = 6.11$$

$$\widehat{(\Phi\Upsilon)}_{33} = +7.66 - 6.11 = 1.55$$

Also, according to Table 8.25, sum of squares explained (or unexplained SS removed)

by temperature is 7666.89
by chemical compounds is 21754.9
by interaction of temperature and compound is 30290.7 − 7666.89 − 21754.9 = 868.9

To test the hypothesis H_0: $\Phi_1 = \Phi_2 = \Phi_3 = 0$ (the temperature effects are all 0); against H_1: H_0 not true; the F ratio [by (4.3)] is

$$F = \frac{(7666.89)/2}{971.313/18} = 71.04$$

Because $F_{.01}$ with $\nu_1 = 2$ and $\nu_2 = 18$ is 6.01, the null hypothesis would have been rejected at the .01 level of significance.

Similarly, the hypothesis H_0: $\Upsilon_1 = \Upsilon_2 = \Upsilon_3 = 0$ (chemical compound effects are all zero); against H_1: H_0 not true; is tested by

$$F = \frac{21754.9/2}{971.313/18} = 201.58$$

with $\nu_1 = 2$, $\nu_2 = 18$, whereas H_0: $(\Phi\Upsilon)_{lj} = 0$; $l = 1, \ldots, f$, $j = 1, \ldots, k$ (interaction effects are all zero); against H_1: H_0 not true; is tested by

$$F = \frac{868.9/4}{971.313/18} = 4.03$$

with $\nu_1 = 4$, $\nu_2 = 18$. The interactions would not have been found significant at the .01 level but would have been had the .05 level been prespecified. ◀

The degrees of freedom for the three components of the regression sum of squares can easily be stated in terms of f, the number of levels of factor 1 (rows), in terms of k, the number of factor 2 (columns), and in terms of r, the number of replications. Table 8.26 shows the breakdown.

Table 8.26 Degrees of Freedom for a Two-Factor Design with Interactions

Sum of Squares due to	df
Rows (factor 1)	$f - 1$
Columns (factor 2)	$k - 1$
Interactions	$(f - 1)(k - 1)$
Error	$fk(r - 1)$
Total	$rfk - 1$

The degrees of freedom for the factors and for the interactions are, of course, equal to the number of regression variables used to represent them. The sum of squares due to error always has $n - m$ degrees of freedom. Here $n = rfk$, the total number of observations; $m = f - 1 + k - 1 +$

$(f - 1)(k - 1) + 1 = fk$, the total number of parameters used in the regression (including β_0).

Note that the sum of squares due to error is zero when $r = 1$. Hence interactions cannot be included in this model when $r = 1$. The total degrees of freedom is always $n - 1$ (or it could have been obtained by addition).

If there are missing observations (r is not a constant), the degrees of freedom can be found by using regression principles as above. The degrees of freedom of the sum of squares due to error will be $n_T - fk$ and the degrees of freedom on the total sum of squares will be $n_T - 1$, where n_T is the total number of observations. Result 8.27, however, will no longer hold. The reader can verify this by dropping an observation in Table 8.24 and calculating the correlations.

As a result, the "contribution" of factor 1, say, to the regression sum of squares now depends on whether factor 1 is entered first, after factor 2, or after factor 2 and the interactions (review Section 8.2). Tests for the significance of factors or interactions may be (but not necessarily[1]) done by using the increase in the regression sum of squares caused by adding that factor or interactions last. The sums of squares thus attributed to factors and interaction do not add up to the sum of squares due to the regression as a whole.

When missing data occur to the point that one or more cells are empty, the interaction effects cannot be found unambiguously. There is then simply not enough data to disentangle all of the interaction effects. Regression coding according to (8.26) will result in extreme multicollinearity. As the reader will verify for him/herself in Exercise 8.20, some tricky coding will permit testing for interactions in general and for certain contrasts as well. A rather comprehensive treatment of the problems of unbalanced data is given in [9], Chapters 6, 7, and 8.

There is more difficulty with missing observations in ANOVA programs (not using regression). If only one or two observations are missing, artificial observations may be entered as mentioned for the randomized block design. When r is a constant, an ANOVA program will give the same results, although generally fewer details, as the regression analysis.

▶ **Example 8.19** The design in Table 8.21 was analyzed by ANVAR4 [8], a time-sharing program. The printout is shown in Table 8.27. ◀

The principle of interactions is not restricted to factorial studies. Designs utilizing randomized block or latin squares may have interaction terms (Exercise 8.8). Also, if three or more factors are used, three or more factors

[1] For a discussion of such matters see J. E. Overall and D. K. Spiegel, "Concerning Least Squares Analysis of Experimental Data," *Psychological Bulletin*, **72**, pp. 311–322, 1969.

may interact (Exercise 8.21). The regression method of coding interactions in these more complex designs still follows the principles outlined in this section.

Table 8.27 ANOVA Printout for the Design in Table 8.21

SOURCE	SUM OF SQUARES	DF	MEAN SQUARE	*F* RATIO
ROWS	7666.87	2	3833.44	71.0398
COLS	21754.9	2	10877.4	201.577
INTER	868.937	4	217.234	4.02571
SUBTOT	30290.7	8		
WITHIN	971.312	18	53.9618	
TOTAL	31262.	26		

PROBABILITY OF $F >= 71.0398$ WITH 2 AND 18 DF IS 0

PROBABILITY OF $F >= 201.577$ WITH 2 AND 18 DF IS 0

PROBABILITY OF $F >= 4.02571$ WITH 4 AND 18 DF IS 1.66907E-02

8.9 ESTIMATION IN THE TWO-FACTOR ANOVA MODEL

If there are no missing variables, estimation of effects directly from the results of the experiment is straightforward and intuitive. Let

$$\begin{aligned} \mu_{\Phi_l} &= \mu + \Phi_l \\ \mu_{\Upsilon_j} &= \mu + \Upsilon_j \\ \mu_{lj} &= \mu + \Phi_l + \Upsilon_j + (\Phi\Upsilon)_{lj} \end{aligned} \tag{8.28}$$

The constant μ_{Φ_l} is the expected value of the dependent variable Y when factor 1, level l, is operating and when the effects of the other factor and of interaction are either zero or have been canceled out in some way; μ_{Υ_j} has a similar definition, but for factor 2. The constant μ_{lj} is the expected value of Y under treatment (l, j).

▶ **Example 8.20** Table 8.28 reproduces Table 8.21 with the addition of averages.

Table 8.28 Effects of Compounds and Curing Temperatures on Adhesive Strength

	Compounds			
Temperature	1	2	3	Average
1	138, 130, 135 av = 134.33	154, 139, 160 av = 151.00	81, 70, 75 av = 75.33	120.22
2	81, 93, 79 av = 84.33	111, 100, 104 av = 105.00	49, 39, 55 av = 47.67	79.00
3	95, 100, 93 av = 96.00	141, 137, 128 av = 135.33	51, 63, 72 av = 62.00	97.78
Average	104.89	130.44	61.67	99.00

The average 120.22 for temperature 1 is uninfluenced by compounds or by interactions because each of these sum to zero over the row. Hence it is reasonable that (and it could be proved)

$$\hat{\mu}_{\Phi_1} = 120.22 = \hat{\mu} + \hat{\Phi}_1$$

Similarly,

$$\hat{\mu}_{\Phi_2} = 79.00 = \hat{\mu} + \hat{\Phi}_2$$

$$\hat{\mu}_{\Phi_3} = 97.78 = \hat{\mu} + \hat{\Phi}_3$$

Adding the three equations and using the condition $\Sigma\Phi_l = 0$

$$\hat{\mu} = \frac{120.22 + 79.00 + 97.78}{3}$$

$$= 99.00$$

Hence

$$\hat{\Phi}_1 = 120.22 - 99.00 = 21.22$$

$$\hat{\Phi}_2 = 79.00 - 99.00 = -20.00$$

$$\hat{\Phi}_3 = 97.78 - 99.00 = -1.22$$

The effects of compounds can be obtained by solving

$$\hat{\mu} + \hat{\Upsilon}_1 = 104.89$$

$$\hat{\mu} + \hat{\Upsilon}_2 = 130.44$$

$$\hat{\mu} + \hat{\Upsilon}_3 = 61.67$$

It seems reasonable (and again could be proved) that μ_{11} be estimated by the average adhesive strength for treatment (1, 1). Hence

$$\hat{\mu}_{11} = 134.33 = \hat{\mu} + \hat{\Phi}_1 + \hat{\Upsilon}_1 + (\hat{\Phi\Upsilon})_{11}$$

Solving,

$$(\hat{\Phi\Upsilon})_{11} = 134.33 - 99.00 - 21.22 - 5.89 = 8.22$$

Similarly,

$$\hat{\mu}_{12} = 151.00 = \hat{\mu} + \hat{\Phi}_1 + \hat{\Upsilon}_2 + (\hat{\Phi\Upsilon})_{12} \quad \text{etc.}$$

Naturally, these estimates are identical to those in Example 8.18. ◀

Contrasts may be estimated for μ_{Φ_l}'s, μ_{Υ_j}'s, or μ_{lj}'s.

$$L = \sum_{l=1}^{f} c_l \mu_{\Phi_l} \qquad \text{where } \sum c_l = 0 \tag{8.29}$$

or

$$L = \sum_{j=1}^{k} c_j \mu_{\Upsilon_j} \qquad \text{where } \sum c_j = 0 \tag{8.30}$$

or

$$L = \sum_{l=1}^{f} \sum_{j=1}^{k} c_{lj} \mu_{lj} \qquad \text{where } \sum\sum c_{lj} = 0 \tag{8.31}$$

Although the estimated variances for the estimators of these contrasts may be obtained from the variance-covariance matrix of regression coefficients in the usual way, the following formulas are often handy:

$$S^2_{\hat{\mu}\Phi_l} = \frac{\text{MSE}}{rk} \tag{8.32}$$

$$S^2_{\hat{\mu}\Upsilon_j} = \frac{\text{MSE}}{rf} \tag{8.33}$$

$$S^2_{\hat{\mu}_{lj}} = \frac{\text{MSE}}{r} \tag{8.34}$$

These estimated variances were derived in a manner similar to the derivation on (7.7). Notice that the divisor of MSE is always the number of observations used to estimate the mean in question.

For a priori estimation of a single contrast

$$t = \frac{\hat{L} - L}{S_{\hat{L}}} \tag{8.35}$$

where t, as usual, has the degrees of freedom of the residual sum of squares.

In the Scheffé simultaneous intervals method

$$SC_\alpha = \sqrt{\nu_1 F(\alpha;\, \nu_1, \nu_2)} \tag{8.36}$$

where $(1 - \alpha) \times 100$ is the percent confidence level of the family of all contrasts on L;

ν_1 is the number of regression variables used to establish the effects that appear in the contrasts in the family; for contrasts on μ_{Φ_i}'s, $\nu_1 = f - 1$; for contrasts on μ_{Υ_j}'s, $\nu_1 = k - 1$; for μ_{ij}'s, $\nu_1 = f - 1 + k - 1 + (f - 1)(k - 1) = fk - 1$;

ν_2 is the degrees of freedom for the residual sum of squares ($\nu_2 = fk(r - 1)$ for a balanced design);

$\Pr(F \geqslant F(\alpha;\, \nu_1, \nu_2)) = \alpha$ where the F distribution has ν_1 and ν_2 degrees of freedom.

These produce the intervals

$$\widehat{SI}_{1-\alpha} = [\hat{L} \pm S_{\hat{L}} SC_\alpha] \tag{8.37}$$

▶ **Example 8.21** The experiment in Example 8.17 was intended to give a 95% confidence interval for the difference in the effects of compounds 3 and 2 on adhesive strength. The point estimate for the required contrast is

$$\begin{aligned}
\hat{L} &= \hat{\mu}_{\Upsilon_3} - \hat{\mu}_{\Upsilon_2} \\
&= \hat{\mu} + \hat{\Upsilon}_3 - (\hat{\mu} + \hat{\Upsilon}_2) \\
&= \hat{\Upsilon}_3 - \hat{\Upsilon}_2
\end{aligned}$$

By Table 8.25

$$\begin{aligned}\hat{L} &= (b_0 - b_3 - b_4) - (b_0 + b_4) \\ &= -b_3 - 2b_4 \\ &= -5.89 - 2(31.44) \\ &= -68.77\end{aligned}$$

The variance of $\hat{L}$ can be obtained with the help of the variance-covariance matrix of regression coefficients from the regression run with the data in Table 8.24 (see Table 8.29).

Table 8.29 Variance-Covariance Matrix of Regression Coefficients for a Run with the Data in Table 8.24

b_1	3.997								
b_2	− 1.999	3.997							
b_3	0	0	3.997						
b_4	0	0	− 1.999	3.997					
b_5	0	0	0	0	7.994				
b_6	0	0	0	0	− 3.997	7.994			
b_7	0	0	0	0	− 3.997	1.999	7.994		
b_8	0	0	0	0	1.999	− 3.997	− 3.997	7.994	
b_0	0	0	0	0	0	0	0	0	1.999
	b_1	b_2	b_3	b_4	b_5	b_6	b_7	b_8	b_0

Using (1.23), we obtain

$$\begin{aligned}S_{\hat{L}}^2 &= S_{b_3}^2 + 4S_{b_4}^2 + 4S_{b_3b_4} \\ &= 3.997 + 4(3.997) + 4(-1.999) \\ &= 11.99\end{aligned}$$

or, alternatively,

$$S_{\hat{L}}^2 = S_{\hat{\mu}_{T_3}}^2 + S_{\hat{\mu}_{T_2}}^2$$

By (8.32)

$$\begin{aligned}S_{\hat{L}}^2 &= \frac{\text{MSE}}{rf} + \frac{\text{MSE}}{rf} \\ &= \frac{2(53.962)}{3(3)} \\ &= 11.99 \\ S_{\hat{L}} &= 3.46\end{aligned}$$

The required confidence interval is

$$\hat{I}_{.95} = [-68.77 \pm 3.46 t_{.025}]$$

Because $\nu = 18$ (Table 8.25),

$$\begin{aligned}\hat{I}_{.95} &= [-68.77 \pm 3.46(2.101)] \\ &= [-76.04, -61.50]\end{aligned}$$

Actually the estimate of the difference in the additive effects of compounds 3 and 2 on adhesive strength is not likely to have much practical significance because of the presence of interaction. Changing from compound 2 to 3 produces different changes in adhesive strengths at different curing temperatures. ◀

▶ **Example 8.22** After inspecting Table 8.28 it was decided to estimate the expected difference in adhesive strengths of treatments (1, 2) and (2, 3) at the 95% confidence level. The estimate for the difference is

$$\begin{aligned}\hat{L} &= \hat{\mu}_{12} - \hat{\mu}_{23} \\ &= \hat{\mu} + \hat{\Phi}_1 + \hat{\Upsilon}_2 + (\hat{\Phi\Upsilon})_{12} - (\hat{\mu} + \hat{\Phi}_2 + \hat{\Upsilon}_3 + (\hat{\Phi\Upsilon})_{23}) \\ &= \hat{\Phi}_1 - \hat{\Phi}_2 + \hat{\Upsilon}_2 - (-\hat{\Upsilon}_1 - \hat{\Upsilon}_2) + (\hat{\Phi\Upsilon})_{12} - (-(\hat{\Phi\Upsilon})_{21} - (\hat{\Phi\Upsilon})_{22}) \\ &= \hat{\Phi}_1 - \hat{\Phi}_2 + \hat{\Upsilon}_1 + 2\hat{\Upsilon}_2 + (\hat{\Phi\Upsilon})_{12} + (\hat{\Phi\Upsilon})_{21} + (\hat{\Phi\Upsilon})_{22}\end{aligned}$$

The point estimate can now be obtained by filling in the point estimates for the individual effects (found in Example 8.18).

Because

$$\hat{L} = b_1 - b_2 + b_3 + 2b_4 + b_6 + b_7 + b_8$$

the estimated variance $S_{\hat{L}}^2$ can be found by using (1.23) and the variance-covariance matrix of regression coefficients in Table 8.29.

However, it is easier to find the point estimate for $\hat{L}$ simply by taking the difference in the averages for treatments (1, 2) and (2, 3) in Table 8.28.

$$\begin{aligned}\hat{L} &= 151.00 - 47.67 \\ &= 103.33\end{aligned}$$

Also, by (8.34) and because $\hat{\mu}_{12}$ and $\hat{\mu}_{23}$ are statistically independent,

$$S_{\hat{L}}^2 = S_{\hat{\mu}_{12}}^2 + S_{\hat{\mu}_{23}}^2$$

$$= \frac{\text{MSE}}{3} + \frac{\text{MSE}}{3}$$

$$= \tfrac{2}{3}(53.962)$$

$$= 35.57$$

$$S_{\hat{L}} = 5.96$$

By (8.36)

$$SC_{.05} = \sqrt{\nu_1 F(.05;\, \nu_1,\, \nu_2)}$$

where

$$\nu_1 = fk - 1 = 8,$$

$$\nu_2 = 9(2) = 18.$$

Therefore

$$SC_{.05} = \sqrt{8F(.05;\, 8,\, 18)}$$

$$= \sqrt{8(2.51)}$$

$$= 4.48$$

so that

$$\widehat{SI}_{.05} = [\,103.33 \pm 5.96(4.48)\,]$$

$$= [\,76.63,\, 130.03\,]$$

Note that other intervals could have been added (by using $SC_{.05} = 4.48$ to calculate them) and the *family* would have had a confidence level of .95

◀

8.10 MODEL APPROPRIATENESS[1]

In the models discussed in the preceding sections various forms of linear effects were included. In a fairly complex design, say, in one using several latin squares, or in a design with two or more factors and perhaps blocks as well, we are often tempted to do extensive model respecification after a

[1] A more complete treatment of this topic can be found in [6], Chapter 10.

preliminary analysis. Interaction terms, carry-over terms, or blocking variables that are not "significant" may be quietly dropped.

When interaction terms are dropped from the model, this procedure is generally called *pooling* the interaction sum of squares with the error sum of squares. (When the interaction effects are dropped from the model, the residual sum of squares is increased by the sum of squares due to interaction.) Interaction effects may use up quite a few degrees of freedom: a two-factor experiment with seven levels for each factor will have 36 degrees of freedom to return to the error term as a result of pooling. The F value needed to demonstrate significance for main factor level effects will thus decrease as a result of pooling. Factor level effects may suddenly become "significant" after pooling.

Unfortunately, pooling and other *post hoc* model modifications are of dubious validity. The old sin of choosing hypotheses on the basis of the data to be analyzed (review Section 4.5) is committed. The consequences of this hypothesis hunting have not been dealt with in theory for this case so neatly as they have been for contrasts (e.g., the Scheffé method of multiple comparisons). Although some rules of thumb for decreasing the dangers of pooling have been suggested ([6], page 126), it is probably best to avoid pooling if possible. If the experiment has enough observations, the degrees of freedom made available by pooling are not needed. Also, the model should be chosen carefully.

In the ANOVA models discussed in the last two chapters the effects, whether they were due to factor levels, blocks, or interactions, were assumed to be additive; for example, a factor level such as "type of advertisement" added or subtracted from sales but did not cause sales to be multiplied or divided by some number. The appropriateness of this assumption is often difficult to judge in advance, and the exact consequences on the validity of the analysis caused by an incorrect assumption of additivity are difficult to determine. Aside from the cynical reason for assuming additivity, namely that additive models are easier to analyze, we can argue that additivity is often a good approximation to nonadditive effects. It has been suggested [10] that this approximation is reasonable if the differences among treatment effects do not exceed 20% of the overall mean μ.

All the ANOVA models discussed assumed a normally distributed error term ϵ that had constant variance. Because it is inevitable that reality will deviate from the model, the consequences of deviations from normality and constant variance must be examined.

Let us first consider the case in which ϵ is not normal. The distribution of ϵ for ANOVA models, as for more general regression models, must be studied through the residuals. The residuals may be plotted as a frequency distribution for the design as a whole and for each treatment separately. Fortunately only extreme deviations from normality affect the validity of tests and estimates. In general, t tests, F tests, t intervals, and Scheffé intervals are robust and are not much affected by lack of normality; for example, it has been suggested [11] that for data with markedly nonnormal residuals $F_{.025}$, when used instead of $F_{.05}$, results in a safe five % significance level. In exceptional cases data may be transformed in order to produce a more normal error term. Ranked data, in which judges may assign the numbers 1, . . . , n to n objects, is such a case. Reference [12] has tables for this kind of transformation.

The basic model assumes that errors are statistically independent, but correlation may occur, for example, when the observations are those of an interviewer whose attitude changes with successive interviews. In another example an individual may be used as a block and successive treatments may cause carry-over effects. The influence of this correlation on tests and estimates depends on its nature and degree (review Sections 6.5, 6.6, and 6.7). In experimental studies randomization may help to remove correlation among errors.

The final departure from the model discussed here is *heterogeneiety of variance*. This condition occurs when the variance of the error term is not the same for each treatment; for example, the variance among observations for a treatment often increases with the magnitude of the treatment mean. Formal tests for equality of variance exist ([3], Section 15.6).

In general, unequal variances have little influence on F tests for the equality of effects if the numbers of replications per cell are all equal. Equally robust is the F-based Scheffé multiple-comparison procedure when cell replications are equal. However, t tests and intervals for treatment means may be seriously affected; for example, suppose that a single-factor design has four factor levels; the first two have large variances among observations, whereas the last two have small variances. Obviously, use of the MSE would result in a confidence interval too narrow for the difference between the first two factor-level means and too wide for the difference in the last two. With serious inequalities in cell replications even the F test and the Scheffé procedure may be inaccurate for unequal variances.

Transformations are the usual cure prescribed for unequal variances; for example, if the error variance is proportional to the treatment mean μ_{ij}, the transformation $Y' = \sqrt{Y}$ is used. As another example, if the observations are percentages, by elementary probability theory the variances will be functions of percentages, hence not constant. The transformation suggested in this case is $Y' = \arcsin \sqrt{\rho}$, where ρ is the observation as a proportion. It can be shown[1] that these transformations tend to equalize variances.

We should remember, however, that in using transformations we are changing models in midstream and consequently may get a little wet. If there really are additive effects on Y, transformations may destroy this additivity. Unbiased estimates for treatment means with the transformed data may not transform back into unbiased estimates of treatment means with the original data. Testing the equality of factor-level means in the single-factor model is no problem, for if the transformations of the means are equal the means are equal. In the presence of interactions, however, equality of means with the transformed model does not guarantee equality of means in the original data.

EXERCISES

8.1 A firm devised a new training method for its machine-parts inspectors. Six candidates for the position were chosen so that there were three groups of two: good technical background, medium technical background, and poor technical background. One member of each group was randomly assigned to the old training method and one to the new. After training all six candidates were given a standard test scored out of 100. The results are given in Table 8.30.

Table 8.30 Scores in a Test of Training Methods

Group background	Old method	New method
Very good	85	85
Medium	60	70
Poor	35	55

(a) Formulate the ANOVA model that you would use to analyze the problem. Define the variables and state the assumptions (very carefully).
(b) Give the regression model that you would use to estimate the parameters in (a) and construct the data matrix for the regression. Specify which regression coefficients estimate which parameters.
(c) Run the regression on an appropriate program and obtain estimates for the parameters specified in (a).
(d) Test the null hypothesis that the new method is the same as the old method against the alternative that the new method is better. Use the .01 level of significance.

[1] See Appendix A, Note 8.2.

8.2 Using the results of the regression run in Exercise 8.1, test the null hypothesis that there is no difference in the effectiveness of the two methods. Use the .01 level of significance. Test the same hypothesis by means of an ANOVA program.

8.3 A randomized block experiment with four factor levels and three blocks yielded the results in Table 8.31:

Table 8.31 A Randomized Block Experiment

	Factor levels			
Blocks	F_1	F_2	F_3	F_4
B_1	85	68	73	75
B_2	76	60	51	60
B_3	83	73	69	70

Use regression runs to complete (a), (b) and (c).

(a) Estimate the effects in the model.
(b) Test the null hypothesis that the factor-level means are all equal at the .05 level of significance.
(c) Simultaneously find 95% confidence intervals for all pairwise differences among the first three factor-level means (Scheffé's method). Use the differences in an attempt to rank the first three factor-level means (of F_1, F_2, and F_3).
(d) Check the estimates in (a) by manual calculation of averages in Table 8.31.
(e) Check the results of (b) by running an ANOVA program.

8.4
(a) Repeat Exercise 8.1, parts (c) and (d), this time assuming that the score for the person with a "very good" background who was taught by the old method is missing.
(b) Obtain the estimates of the effects in Exercise 8.1 (with the observation missing as above) by calculations taken directly from Table 8.30. Check the answer with part (a). This question is a bit tricky.

8.5 The experiment in Exercise 8.3 is to be repeated. The number of replications per cell necessary to produce confidence intervals within ± 5 in the differences observed, at the 95% confidence level, is required. Use the results of Exercise 8.3.

8.6 The experiment in Exercise 8.3 is to be repeated. The difference in the averages $(\mu_{F_1} + \mu_{F_2})/2$ and $(\mu_{F_3} + \mu_{F_4})/2$ must be estimated within ± 5 at the 99% level of confidence. Find the number of replications per cell required. Use the results of Exercise 8.3.

8.7 "In planning sample size for a randomized block experiment, we guessed a value of 14.36 for the MSE (mean square attributed to error). Manipulation of formulas for the intervals showed that the most precise results for any sample size would be obtained if we used no blocks at all." Comment.

8.8 In the situation of Example 8.1 the experiment was repeated, except that 24 people (different ones) were now used as in Table 8.32.

Table 8.32 Scores on a Computational Efficience Test

Mathematical sophistication	Calculator 1	2	3(Own)
1	80, 86	91, 93	74, 81
2	74, 69	83, 79	61, 75
3	63, 79	71, 76	59, 63
4	61, 71	65, 71	68, 68

(a) Estimate the block and calculator effects according to model (8.1) by using regression.
(b) Check the estimated effects found in (a) by manual computation of the appropriate averages in Table 8.32.
(c) Perform a lack of fit analysis (see Section 5.5) and test for lack of fit at the .05 level of significance.
(d) Estimate the difference between the expected score of calculator 3 and the average expected score of calculators 1 and 2 at the .05 level of significance. Assume that the estimate was suggested by the data.

8.9 Consider the results of the following latin square experiment (Table 8.33).

Table 8.33 Results of a Latin Square Experiment; Letters Represent Treatments

Blocking variable 2		Blocking variable 1 Blocks 1	2	3	4
	1	A, 63	B, 49	C, 17	D, 54
Blocks:	2	B, 72	A, 56	D, 39	C, 59
	3	C, 54	D, 40	A. 39	B, 65
	4	D, 50	C, 36	B, 40	A, 63

(a) Construct a data matrix for the corresponding regression model.
(b) (1) Obtain estimates for block and treatment effects from a regression run; (2) verify these estimates by calculation from appropriate averages in Table 8.33.
(c) Use regression results to test the hypothesis that the effects of treatments A, B, C, and D are all zero. Use the .05 level of significance.

(d) Use regression results to construct an analysis of variance table that attributes sums of squares to each of the blocking variables and to the factor levels (treatments).
(e) Check the results of part (c) by running an ANOVA program; compare with part (d).
(f) Obtain a 95% Scheffé interval for the difference between the average of effects A and B and the average of effects C and D.
(g) Repeat parts (a), (b) 1, (c), and (d) under the assumption that the observation B, 49 is missing.

8.10 Consider the experimental results of Table 8.18 in Example 8.14.

(a) Formulate an ANOVA model that contains effects due to each of the six stores, time periods, and potato grades.
(b) Formulate the model in part (a) as a regression model. The effects of the first five stores should appear directly as coefficients in the regression. Construct a regression data matrix, run the regression, and obtain estimates for the effects.
(c) Check the estimates for the effects in (b) by computing the appropriate averages from Table 8.18.
(d) Use the results of part (c) to test the hypothesis (at the .05 level of significance) that there is a difference between urban and suburban stores. *Hint*: Test the hypothesis that the difference in average effects of the two types of store is zero. Assume an *a priori* test.
(e) Test the hypothesis that potato grades have no effect on sales. Use the .01 level of significance.

8.11
(a) Repeat Exercise 8.10, parts (a) and (b); this time the effects of the urban stores should sum to zero and the effects of the suburban effects should sum to zero. An effect due to "urbanity" and one due to "suburbanity" should be present; these two effects should sum to zero. Show that this model is actually equivalent to the one in part (a), Exercise 8.10.
(b) Use the results of part (a) to test the hypothesis (at the .05 level of significance) that there is a difference between urban and suburban stores. *Hint*: The regression will have a variable that distinguishes urban and suburban stores; use an F test on this variable. Show that the t test in Exercise 8.10 part (d) is equivalent to the F test.

8.12 Consider the experimental results in Table 8.18, Example 8.14.

(a) Formulate an ANOVA model, as in Exercise 8.10 or 8.11, but also include additive carry-over effects as specified by the small letters in Table 8.18.
(b) Formulate the model in part (a) as a regression model.
Hint: Use

$$X_a = \begin{cases} 1 & \text{when } a \text{ is present} \\ -1 & \text{when } c \text{ is present} \\ 0 & \text{otherwise;} \end{cases} \qquad X_b = \begin{cases} 1 & \text{when } b \text{ is present} \\ -1 & \text{when } c \text{ is present} \\ 0 & \text{otherwise.} \end{cases}$$

Run the regression and obtain estimates for all effects.

(c) Test the hypothesis that there are no carry-over effects at the .05 level of significance.
(d) Test the hypothesis that potato grades have no effect on sales. Use the .01 level of significance.
(e) A scheme for coding the carry-over effects alternative to the one suggested in part (b) is

$$
\begin{aligned}
X_a &= 1 && \text{when } a \text{ is present}\\
&= -1 && \text{when no carry-over exists}\\
&= 0 && \text{otherwise,}\\
X_b &= 1 && \text{when } b \text{ is present}\\
&= -1 && \text{when no carry-over exists}\\
&= 0 && \text{otherwise,}\\
X_c &= 1 && \text{when } c \text{ is present}\\
&= -1 && \text{when no carry-over exists}\\
&= 0 && \text{otherwise.}
\end{aligned}
$$

The scheme is incorrect. Why?

8.13 It was decided to use average sales for the five weeks preceding the experiment in each store in Example 7.1 as a concomitant variable. Table 8.34 is Table 7.1 with the addition of this concomitant variable.

Table 8.34 Sales of Potatoes in Number of Bags per Week with Average Previous Sales per Store

Design 1	Average previous sales of store	Design 2	Average previous sales of store	Design 3	Average previous sales of store
52	24	23	20	21	21
36	19	36	23	32	22
24	21	29	21	17	19
48	22	16	20	24	24
52	22	27	21	15	19
27	19				
48	22				
41	20				

(a) Formulate a covariance ANOVA model with previous sales as a concomitant variable.
(b) Formulate the model in part (a) as a regression model and obtain estimates for all parameters.
(c) Test the hypothesis that all three designs produce equivalent sales (at the .01 level of significance).
(d) Use Scheffé intervals at the 95% confidence level in an attempt to rank the three designs.

8.14 The experimental results of Table 8.34 in Exercise 8.13 were analyzed by ANCOV [13], a time-sharing program. The outcome is in Table 8.35. Use the results of the regression in Exercise 8.13 to explain every line in the printout except the "SUM SQRS X" and "SUM XY."

Table 8.35 Printout of Program ANCOV for the Data in Table 8.34

	BETWEEN	THIN	TOTAL
DF	2	15	17
SUM SQRS X	6.93359E-02	44.875	44.9443
SUM XY	9.44531	169	178.445
SUM SQRS Y	1332.84	1247.6	2580.45
ADJ SS Y	1260.81	611.145	1871.95
ADJ DF	2	14	16
MEAN SQR	630.404	43.6532	116.997
F	14.4412		
MEAN ADJ Y(1)	40.7385		
MEAN ADJ Y(2)	26.4092		
MEAN ADJ Y(3)	22.0092		

8.15 A contractor involved in large-scale construction projects around the world ran a continuing program in cost estimation for its employees. More than one-half of the 200 employees involved in cost estimation had taken the course. Eighteen employees were chosen to evaluate a special project as a test of the effectiveness of the course. Their estimates are given in Table 8.36.

Table 8.36 Project Cost Estimates in Thousands of Dollars

	Years of experience		
	Zero up to 5 years	5 to 10 years	more than 10 years
Had taken course	143	149	146
	145	139	133
	133	142	139
Had not taken course	99	76	117
	112	85	105
	117	87	110

(a) Formulate a model for the problem (including interactions) and estimate the parameters of the model by using linear regression. Check the results by estimating directly from Table 8.36.

(b) Test the hypothesis that the course does not make a difference in cost estimates (.01 level of significance). Test the hypothesis that experience does not make a difference in cost estimates (.05 level of significance). Test the hypothesis that experience and the presence of the course do not interact (at the .05 level of significance).

(c) Run an appropriate ANOVA program and check the results of part (b).

(d) Estimate the difference in the cost estimates of subjects who have taken the course and have 0 to 5 years experience and subjects who have not taken the course and have 5 to 10 years experience. Use the 95% confidence level but assume that this estimate was suggested by the data.

8.16 Three students had taken two courses each in each of mathematics and English literature. Their scores on examinations are given in Table 8.37:

Table 8.37 Scores on Examinations

Student	Mathematics	English literature
1	63, 57	95, 83
2	80, 75	55, 60
3	51, 48	75, 69

It was decided to use an analysis of variance model to analyze the results. A regression utilized the following variables:

X_1 = student one
X_2 = student two
X_3 = mathematics
X_4 = interaction between English Literature and student 2
X_5 = interaction between mathematics and student 3

The regression was designed to give *direct* additive effects due to student number, type of course, and interaction (the effects summed to zero).

(a) Give the data matrix that was used to run the regression.

(b) Using the data matrix of part (a), two regressions were run (Table 8.38).

Construct an analysis of variance table, giving sums of squares, mean squares. and degrees of freedom for row effects, column effects, and interaction effects.

8.17

(a) Repeat Exercise 8.8, part (a), this time assumimg interaction between mathematical sophistication and calculator brands.

(b) Test the hypothesis (using the regression model in part (a)) that there is no interaction at the .05 level of significance.

(c) Construct an analysis of variance table, giving sums of squares, mean squares, and degrees of freedom for sophistication effects, calculator effects, and interaction effects.

Table 8.38 Two Regression Runs

RUN 1

ANALYSIS OF VARIANCE	DF	SUM OF SQUARES	MEAN SQUARE
REGRESSION	2	378.16667	189.08333
RESIDUAL	9	1884.75000	209.41667

VARIABLE	B	STD ERROR B	F / SIGNIFICANCE
X_1	6.9166667	5.9078573	1.3706725 .272
X_2	−.83333333E-01	5.9078573	.19896538E-03 .989
(CONSTANT)	67.583333	4.1774860	261.72742 0.000

VARIANCE/COVARIANCE MATRIX OF THE UNNORMALIZED REGRESSION COEFFICIENTS.

X_1	34.90278	
X_2	− 17.45139	34.90278
	X_1	X_2

RUN 2

ANALYSIS OF VARIANCE	DF	SUM OF SQUARES	MEAN SQUARE
REGRESSION	5	2125.41667	425.08333
RESIDUAL	6	137.50000	22.91667

VAR	B	STD ERROR B	F / SIGNIFICANCE
X_1	6.9166667	1.9543399	12.525455 .012
X_3	− 5.2500000	1.3819270	14.434727 .009
X_4	− 15.250000	1.9543399	60.889091 .000
X_2	−.83333333E-01	1.9543399	.18181818E-02 .967
X_5	− 6.0000000	1.9543399	9.4254545 .022
(CONST)	67.583333	1.3819270	2391.7127 .000

Table 8.38 (continued)

VARIANCE/COVARIANCE MATRIX OF THE UNNORMALIZED REGRESSION COEFFICIENTS.

X_1	3.81944				
X_2	−1.90972	3.81944			
X_3	0.00000	0.00000	1.90972		
X_4	0.00000	0.00000	0.00000	3.81944	
X_5	0.00000	0.00000	0.00000	1.90972	3.81944
	X_1	X_2	X_3	X_4	X_5

(d) If interaction were not significant, would it be a good idea to use a model without interaction terms? Discuss.

(e) Assume that one of the objects of the experiment was to compare the scores of the firm's own calculator with the average of the other two at the fourth level of sophistication (which formed most of the market potential). How large would r have to be if this difference were to be estimated within ± 3 at the 95% confidence level. This estimate was to be made with several others simultaneously.

8.18 To the regression model in Exercise 8.12 add interaction effects between the type of store and time periods. (*Hint*: two extra variables are required.)

(a) Give reasons why such interactions might exist. What other interactions are possible?

(b) Run the regression and test the hypothesis that all the grades of potatoes produce the same effect (.05 level of significance).

(c) Construct an analysis of variance table giving sums of squares, mean squares, and degrees of freedom for store effects, "urbanity" effects, period effects, treatment effects, carry-over effects and interaction effects. Do these sums of squares add up to the sums of squares due to regression.?

8.19 A randomized block experiment with missing data yielded the results in Table 8.39.

Table 8.39 A Randomized Block Design with Missing Data

	Factor levels			
Blocks	F_1	F_2	F_3	F_4
B_1	4, 5	7, 8		
B_2	5	7		
B_3			4	3
B_4			9, 8	

The data above is said to be *disconnected* (see Reference [8.9], Section 7.4 for a discussion of connectedness). The data under blocks B_1 and B_2 could be called set #1 and the rest set #2. Note that no block or factor level acts on both sets.

(a) Use the usual regression coding with three dummy variables for the blocking variable and three for the factor. Run the regression. Why is there multicollinearity? Give an "intuitive" explanation. Express one of the independent variables as a function of the others.
(b) Consider the model suggested by Table 8.40.
Construct coding that reflects this model (assume that there is a different overall mean μ for each set). Run the regression and obtain estimates for the additive effects. Why is the multicollinearity not perfect again?
(c) Using the result of the regression run in (b), test the hypothesis $H_0 : \Upsilon_1' = \Upsilon_1 = 0$; against $H_1 : H_0$ not true; at the .05 level of significance. Is this a meaningful test in the practical sense?

Table 8.40 Suggested Model for Table 8.31

		Factor levels			
		F_1	F_2	F_3	F_4
Blocks	Effects	Υ_1	$-\Upsilon_1$	Υ_1'	$-\Upsilon_1'$
B_1	ρ_1	4, 5	7, 8		
B_2	$-\rho_1$	5	7		
B_3	ρ_1'			4	3
B_4	$-\rho_1'$			9, 7	

8.20 A two-factor experiment with missing data gives the results of Table 8.41.

Table 8.41 A Two-Factor Experiment with Missing Data

		Factor 1	
		F_1	F_2
Factor 2	F_1'	2, 4	5
	F_2'	6, 7	
	F_3'	2, 3	3, 4

(a) Show that there is extreme multicollinearity in the standard regression coding [per (8.26)] for a two-factor model with interactions by expressing one of the independent variables as a linear function of the others.
(b) Run the model without the interaction terms and obtain an analysis of variance table for this regression.
(c) The following device can be used to test for the presence of interactions and to estimate certain contrasts. Code the data in Table 8.41 by using a dummy

variable to designate the total effect on each occupied cell but one. The sum of these effects is to equal zero. This model is used to estimate the cell means (the μ_{ij}'s) for each occupied cell without separating the factor and interaction effects. Verify that had all the cells been occupied the number of dummy variables obtained would have been equal to the number of dummy variables found in (a). Actually this method of coding is basically equivalent to (8.26) when all the cells are occupied. For a discussion of the problems with "unbalanced" data see [9], Chapters 6, 7, and 8. Run the regression with this coding and obtain the residual sum of squares. Because this model can be viewed as an expansion of the one in (b) to include interactions, (4.3) can now be used to test for the presence of interactions (while assuming the presence of factor effects). Perform the test at the .05 level of significance.

(d) Estimate the difference $\mu_{11} - \mu_{32}$ at the 95% confidence level, using the models in (b) and (c). Use Scheffé intervals. Express the difference in each case in terms of factor level effects and interaction effects.

8.21 A researcher devised a test that measured the "initial interest" aroused by an advertisement. Scores were recorded on a scale of 0 to 50 by measuring certain physical manifestations of interest in the potential customer. The 24 experimental subjects were divided into income, age, and sex categories. Their scores are given in Table 8.42.

Table 8.42 Interest Scores of 24 Experimental Subjects

		Young	Middle-Aged
Male	High income	24, 8	20, 4
	Medium income	12, 8	32, 36
	Low income	28, 8	16, 4
Female	High income	20, 12	24, 4
	Medium income	20, 12	28, 12
	Low income	32, 36	20, 24

The above could be considered a "three-factor" experiment with income, age, and sex as factors. Assume that there are additive effects due to the three factors and that there are interactions as well. In particular, there could be interactions between each pair of factors (sex-age, sex-income, age-income) and among three factors at a time (sex-age-income).

(a) Construct an ANOVA model incorporating main effect, pairwise interactions, and triple interactions.

(b) Code the model as a regression problem by extending the method in (8.26). *Hint*: You should have

4 dummy variables for main effects
2 dummy variables for income-age interactions
2 dummy variables for income-sex interactions
1 dummy variable for age-sex interactions

2 dummy variables for income-age-sex interactions.

Run the regression and find the effects.

(c) By using the proper regression runs construct an analysis of variance table that separates the various "sources" of the sum of squares due to regression given in the "Hint" above and provide an F value (with the appropriate degrees of freedom) to test the significance of each set of effects. Note that for balanced (no missing values) data regression provides a rather computationally inefficient method of analyzing a balanced three-factor experiment. ANOVA programs utilizing more direct methods are available.

REFERENCES

1. J. L. Mulchy, *ANVAR3*, Hewlett-Packard Time-Shared Basic Program Library.
2. N. R. Draper and H. Smith, *Applied Regression Analysis*. New York: Wiley, 1966.
3. John Neter and William Wasserman, *Applied Linear Statistical Models*; Homewood, Ill.: Irwin, 1974.
4. J. L. Mulchy, *ANVAR2*, Hewlett-Packard Time-Shared Basic Program Library.
5. Peter W. M. John, *Statistical Design and Analysis of Experiments*, New York: Macmillan, 1971.
6. Henry Scheffé, *The Analysis of Variance*. New York: Wiley, 1959.
7. Graduate School of Business of the University of Chicago, *Interactive Data Analysis IDA*, Hewlett-Packard Time-Shared Basic Program Library.
8. J. L. Mulchy, *ANVAR4*, Hewlett-Packard Time-Shared Basic Program Library.
9. S. R. Searle, *Linear Models*. New York: Wiley, 1971.
10. Seymour Banks, *Experimentation in Marketing*. New York: McGraw-Hill, 1965.
11. W. G. Cochran, "Some Consequences when the Assumptions for the Analysis of Variance are not Satisfied," *Biometrics*, **3**, 22–38, March 1947.
12. R. A. Fisher and F. Yates, *Statistical Tables for Biological, Agricultural and Medical Workers*. Edinburgh and London: Oliver & Boyd, 1948.
13. John Ingold, *ANCOV*, Hewlett-Packard Time-Shared Basic Program Library.

appendix A

Mathematical Notes

NOTE 1.1

The integral sign in (1.2) can be thought of as depicting an infinitely large summation of infinitely small parts. Suppose that the range (r_1, r_2) is divided into $(r_2 - r_1)/dX$ intervals of width dX. Let X_i be the midpoint of each interval, as shown in Figure A1.1.

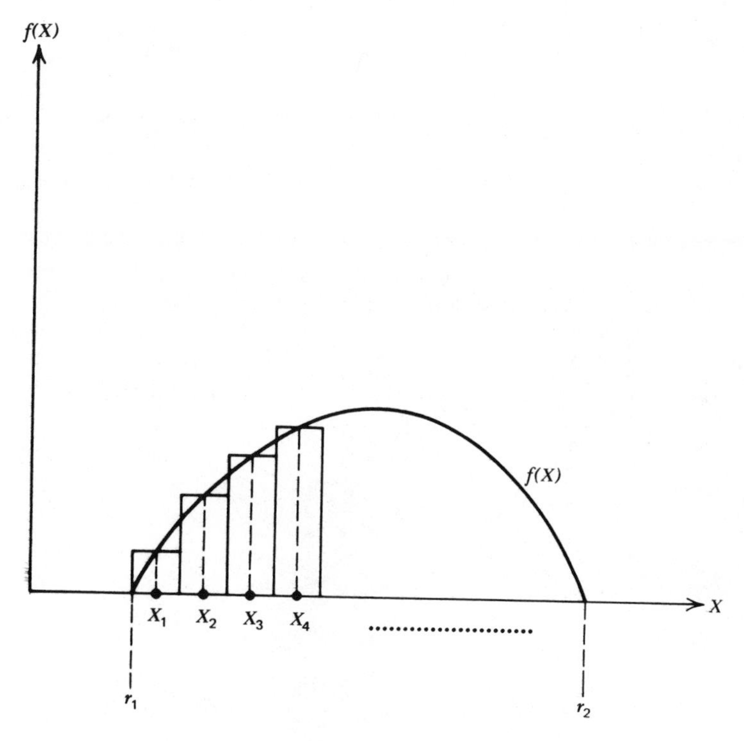

Figure A1.1

Approximate probabilities for $X_1, X_2, \ldots, X_k$ can be calculated in Table A1.1.

Table A1.1

X	X_1	X_2	$\cdots$	X_k
$\Pr(X)$	$f(X_1)\,dX$	$f(X_2)\,dX$	$\cdots$	$f(X_k)\,dX$

where $k = (r_2 - r_1)/dX$. As $dX \to 0$, $k \to \infty$, and $\Sigma_{i=1}^{k} X_i f(X_i)\,dX$ becomes $\int_{r_1}^{r_2} X f(X)\,dX$.

NOTE 1.2

The expected value of $\bar{X}$ is

$$E(\bar{X}) = E\left(\frac{\sum_{i=1}^{n} X_i}{n}\right)$$

$$= \frac{1}{n} E\left(\sum X_i\right)$$

$$= \frac{1}{n}\left[\sum (E(X_i))\right] \qquad \text{(Rule 3, Section 1.3)}$$

$$= \frac{1}{n}[nE(X)]$$

$$= E(X)$$

$$= \mu_X$$

The variance of $\bar{X}$ is

$$\mathrm{VAR}(\bar{X}) = \mathrm{VAR}\left(\frac{\sum_{i=1}^{n} X_i}{n}\right)$$

$$= \frac{1}{n^2}\,\mathrm{VAR}\left(\sum X_i\right) \qquad \text{(Rule 2, Section 1.4)}$$

$$= \frac{1}{n^2} \sum [\mathrm{VAR}(X)]$$

because the X_i's are statistically independent (read Section 1.13);

$$\therefore \quad \mathrm{VAR}(\bar{X}) = \frac{1}{n^2}[n\,\mathrm{VAR}(X)]$$
$$= \frac{\mathrm{VAR}(X)}{n}$$
$$= \frac{\sigma_X^2}{n}$$

NOTE 1.3

$$E((X_1 - E(X_1))(X_2 - E(X_2)))$$
$$= E[X_1X_2 - X_2E(X_1) - X_1E(X_2) + E(X_1)E(X_2)]$$
$$= E(X_1X_2) - E(X_1)\,E(X_2) - E(X_1)\,E(X_2) + E(X_1)\,E(X_1)$$
$$= E(X_1X_2) - E(X_1)\,E(X_2)$$

However, if X_1 and X_2 are statistically independent, $E(X_1X_2) = E(X_1) \cdot E(X_2)$. This can be proved easily for both discrete and continuous random variables. In the continuous case

$$E(X_1X_2) = \int_{-\infty}^{\infty}\int_{-\infty}^{\infty} X_1X_2 f(X_1, X_2)dX_1\,dX_2$$
$$= \int_{-\infty}^{\infty} X_1 f(X_1)\,dX_1 \cdot \int_{-\infty}^{\infty} X_2 f(X_2)\,dX_2 = E(X_1)\cdot E(X_2)$$

Therefore $E((X_1 - E(X_1))(X_2 - E(X_2))) = 0$ if X_1 and X_2 are statistically independent.

NOTE 2.1

If there is a random error in measuring Y, denoted by γ_i, then

$$Y_i + \gamma_i = \beta_0 + \beta_1 X_{1i} + \cdots + \beta_q X_{qi} + \epsilon_i$$

Therefore

$$Y_i = \beta_0 + \beta_1 X_{1i} + \cdots + \beta_q X_{qi} + \epsilon_i - \gamma_i$$

This is the same model with $\epsilon_i' = \epsilon_i - \gamma_i$.

Suppose, however, that there is an error in measuring X_{1i}, denoted by δ_i. Then,

$$Y_i = \beta_0 + \beta_1 (X_{1i} + \delta_i) + \cdots + \beta_q X_{qi} + \epsilon_i$$

or,

$$Y_i = \beta_0 + \beta_1 X_{1i} + \cdots + \beta_q X_{qi} + \epsilon_i + \beta_1 \delta_i$$

The error term $\epsilon_i + \beta_1 \delta_i$ is now a function of one of the parameters of the equation. If such an error term exists and we ignore it by assuming (2.4), it can be shown that we will obtain a biased estimate of β_1.

NOTE 2.2

The problem is to find $b_0, b_1, b_2, \ldots, b_q$ to minimize

$$\sum_{i=1}^{n} (Y_i - (b_0 + b_1 X_{1i} + b_2 X_{2i} + \cdots + b_q X_{qi}))^2$$

or minimize

$$\sum_{i=1}^{n} \epsilon_i^2$$

Set

$$\frac{\partial \sum e_i^2}{\partial b_0} = -\sum_{i=1}^{n} 2\left(Y_i - b_0 - \sum_{j=1}^{q} b_j X_{ji} \right) = 0$$

$$\frac{\partial \sum e_i^2}{\partial b_k} = -\sum_{i=1}^{n} 2\left(Y_i - b_0 - \sum_{j=1}^{q} b_j X_{ji} \right) \cdot X_{ki} = 0, \qquad k = 1, \ldots, q$$

These $q + 1$ equations, called the *normal equations*, can usually be solved to obtain $b_0, b_1, \ldots, b_q$. Matrix algebra provides a neat way of solving these equations: let

$$Y = \begin{bmatrix} Y_1 \\ Y_2 \\ \vdots \\ Y_n \end{bmatrix}, \qquad X = \begin{bmatrix} 1 & X_{11} & \cdots & X_{q1} \\ 1 & X_{12} & \cdots & X_{q2} \\ \vdots & \vdots & & \vdots \\ 1 & X_{1n} & \cdots & X_{qn} \end{bmatrix}, \qquad B = \begin{bmatrix} b_0 \\ b_1 \\ \vdots \\ b_q \end{bmatrix}$$

Then the normal equations can be written as

$$\begin{bmatrix} n & \sum X_{1i} & \sum X_{2i} & \cdots & \sum X_{qi} \\ \sum X_{1i} & \sum X_{1i}X_{1i} & \sum X_{2i}X_{1i} & \cdots & \sum X_{qi}X_{1i} \\ \vdots & \vdots & \vdots & & \vdots \\ \sum X_{qi} & \sum X_{1i}X_{qi} & \sum X_{2i}X_{qi} & \cdots & \sum X_{qi}X_{qi} \end{bmatrix} \begin{bmatrix} b_0 \\ b_1 \\ \vdots \\ b_q \end{bmatrix} = \begin{bmatrix} \sum Y_i \\ \sum X_{1i}Y_i \\ \vdots \\ \sum X_{qi}Y_i \end{bmatrix}$$

or

$$(X'X)B = X'Y$$

$$B = (X'X)^{-1}X'Y$$

If any one of the normal equations is not independent of the others, then $(X'X)^{-1}$ does not exist and the normal equations cannot be solved.

NOTE 2.3

Suppose that we have $(1 - \alpha) \cdot 100\%$ confidence intervals for β_i and β_j, $I_{i,1-\alpha}$ and $I_{j,1-\alpha}$, respectively. The confidence that $I_{i,1-\alpha}$ and $I_{j,1-\alpha}$ are true simultaneously is at least $(1 - 2\alpha) \cdot 100\%$.

Proof Let $\Pr(\bar{I}_{i,1-\alpha} \cap \bar{I}_{j,1-\alpha})$ be the probability that both intervals are incorrect (a priori). Then by basic probability theory

$$\Pr\left(\bar{I}_{i,1-\alpha} \cap \bar{I}_{j,1-\alpha}\right) = \Pr\left(\bar{I}_{i,1-\alpha}\right) + \Pr\left(\bar{I}_{j,1-\alpha}\right) - \Pr\left(\bar{I}_{i,1-\alpha} \cup \bar{I}_{j,1-\alpha}\right)$$

Because $\Pr(\bar{I}_{i,1-\alpha} \cup \bar{I}_{j,1-\alpha})$, the probability that either or both intervals are incorrect, must be greater than 0, then

$$\Pr\left(\bar{I}_{i,1-\alpha} \cap \bar{I}_{j,1-\alpha}\right) \leqslant \Pr\left(\bar{I}_{i,1-\alpha}\right) + \Pr\left(\bar{I}_{j,1-\alpha}\right) = 2\alpha$$

The joint confidence for both intervals is $(1 - 2\alpha) \cdot 100$ *as a lower limit.*

This method of finding joint intervals is called *the Bonferroni method* and can be extended to f intervals simultaneously; that is, the confidence interval would be at least $(1 - f\alpha) \cdot 100$. For example, if four intervals have a 99% confidence level ($\alpha_i = .01$ for $i = 1, 2, 3, 4$), the confidence in the four simultaneously is at least $[1 - 4(.01)] \times 100$, or 96%.

NOTE 2.4

From Note 2.2 (in the expression for the last q normal equations)

$$-\sum_{i=1}^{n} 2\left(Y_i - b_0 - \sum_{j=1}^{q} b_j X_{ji}\right)X_{ki} = 0$$

$$\therefore \quad \sum_{i=1}^{n} e_i X_{ki} = 0 \qquad \text{for } k = 1, \ldots, q$$

This result can be used to prove that $\Sigma(\hat{Y}_{Ri} - \bar{Y})e_i = 0$:

$$\begin{aligned}\sum(\hat{Y}_{Ri} - \bar{Y})e_i &= \sum(b_0 - \bar{Y} + b_1X_{1i} + b_2X_{2i} + \cdots + b_q X_{qi})e_i \\ &= \sum(b_0 - \bar{Y})e_i + 0 \\ &= (b_0 - \bar{Y})\sum e_i \qquad \text{(See Note 2.5)} \\ &= 0\end{aligned}$$

Now

$$Y_i = \hat{Y}_{Ri} + e_i$$

hence

$$\begin{aligned}\sum(Y_i - \bar{Y})^2 &= \sum(\hat{Y}_{Ri} - \bar{Y} + e_i)^2 \\ &= \sum(\hat{Y}_{Ri} - \bar{Y})^2 + \sum 2e_i(\hat{Y}_{Ri} - \bar{Y}) + \sum e_i^2\end{aligned}$$

and because $\sum(\hat{Y}_{Ri} - \bar{Y})e_i = 0$

$$\sum(Y_i - \bar{Y})^2 = \sum(\hat{Y}_{Ri} - \bar{Y})^2 + 0 + \sum(Y_i - \hat{Y}_{Ri})^2$$

NOTE 2.5

From Note 2.2

$$-\sum_{i=1}^{n} 2\left(Y_i - b_0 - \sum_{j=1}^{q} b_j X_{ji}\right) = 0$$

$$\therefore \quad -2\sum e_i = 0$$

$$\therefore \quad \sum e_i = 0 \quad \text{and} \quad \frac{\sum e_i}{n} = \bar{e} = 0$$

NOTE 3.1

When testing H_0: $\beta_1 = 0$ against H_1: $\beta_1 \neq 0$, the critical values of b_1, namely b_1' and b_1'', can be found by solving (from Table 2.3, $S_{b_1} = 1.95$)

$$t_{.005} = 3.012 = \frac{b_1'' - 0}{1.95}$$

and

$$-3.012 = \frac{b_1' - 0}{1.95}$$

to give $b_1'' = 5.87$ and $b_1' = -5.87$ (see Figure A3.1). To find the probability of Type II error when $\beta_1 = 3$ we must find the shaded area in Figure A3.2. From Table 2 we can find that the shaded area is between .9 and .95. Hence the probability of Type II error is greater than 90%.

NOTE 3.2

$$b_j = \frac{\Delta Y}{\Delta X_j} \quad \text{with all other } X_i\text{'s } (i \neq j) \text{ constant}$$

$$= \frac{S_Y\, \Delta Y'}{S_{X_j}\, \Delta X_j'}$$

$$\therefore \quad \frac{\Delta Y'}{\Delta X_j'} = \text{BETA}_j = \frac{b_j S_{X_j}}{S_Y}$$

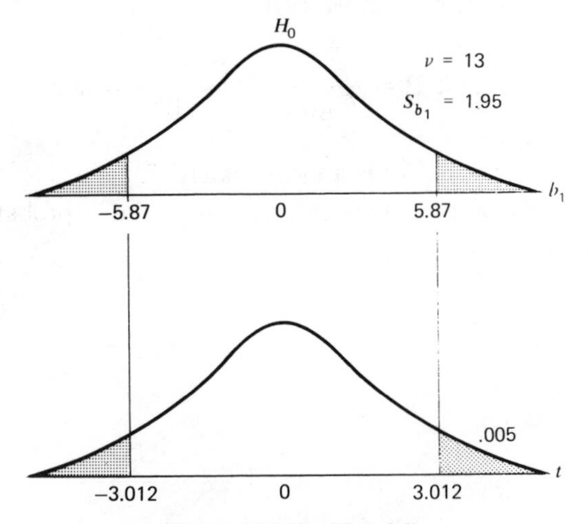

Figure A3.1 (for Note 3.1)

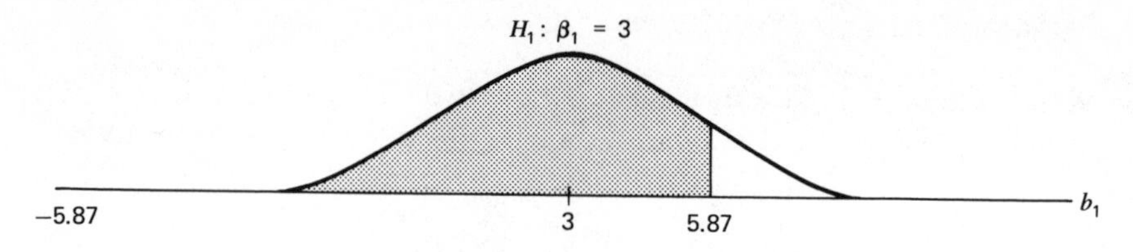

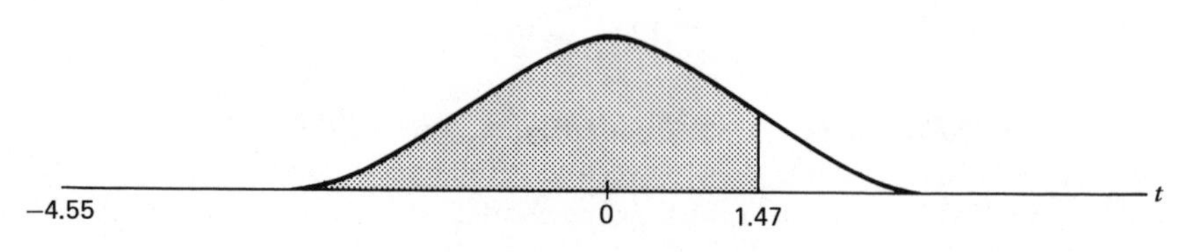

Figure A3.2 (for Note 3.1)

NOTE 3.3

Let $F_p(\nu_1, \nu_2)$ be the F value for a distribution with ν_1 and ν_2 degrees of freedom such that $P_r(F \geqslant F_p(\nu_1, \nu_2)) = p$. It can be shown that

$$F_p(\nu_1, \nu_2) = \frac{1}{F_{1-p}(\nu_2, \nu_1)}$$

From Table 26.9 in the *Handbook of Mathematical Functions*, U.S. Department of Commerce, National Bureau of Standards, Applied Mathematics Series 55, November 1970,

$$F_{.75}(17, 2) = \frac{1}{F_{.25}(2, 17)} = \frac{1}{1.51} = .662$$

Hence, for $F_p(17, 2) = .667$, p is approximately .75.

More accurately, approximate integration of the F probability density gives $p = .749$.

NOTE 4.1

From (4.1)

$$F = (n - m) \cdot \frac{\mathrm{USS}_{Y \cdot q-1;j} - \mathrm{USS}_{Y \cdot q}}{\mathrm{USS}_{Y \cdot q}}$$

dividing numerator and denominator by $\text{USS}_{Y\cdot q-1;j}$

$$F = (n - m) \cdot \frac{(\text{USS}_{Y\cdot q-1;j} - \text{USS}_{Y\cdot q})/\text{USS}_{Y\cdot q-1;j}}{\text{USS}_{Y\cdot q}/\text{USS}_{Y\cdot q-1;j}}$$

$$= (n - m) \cdot \frac{r^2_{Yj\cdot q}}{(\text{USS}_{Y\cdot q-1;j} - \text{USS}_{Y\cdot q-1;j} + \text{USS}_{Y\cdot q})/\text{USS}_{Y\cdot q-1;j}}$$

$$= (n - m) \cdot \frac{r^2_{Yj\cdot q}}{1 - r^2_{Yj\cdot q}}$$

Rearranging,

$$r^2_{Yj\cdot q} = \frac{F}{(n - m) + F}$$

NOTE 4.2

If $\hat{Y}_R = b_0 + b_1X_1 + \cdots + b_qX_q$, then the elasticity of $\hat{Y}_R$ with respect to X_j is

$$E = \lim_{\Delta X_j \to 0} \frac{\Delta \hat{Y}_R / \hat{Y}_R}{\Delta X_j / X_j} \Bigg| X_i \text{ constant for } i \neq j$$

$$= \frac{\partial \hat{Y}_R}{\partial X_j} \cdot \frac{X_j}{\hat{Y}_R}$$

$$= b_j \frac{X_j}{\hat{Y}_R}$$

Because Y_R is a function of $X_1, X_2, \ldots, X_q$, so is E. The approximation for E used in the SPSS program is

$$E' = b_j \frac{\overline{X}_j}{\overline{Y}}$$

NOTE 5.1

If $Y_R = \beta_0 X_1^{\beta_1} X_2^{\beta_2} \cdots X_q^{\beta_q}$, then the elasticity of Y_R with respect to X_j is

$$E_{Y_R X_j} = \frac{\partial Y_R}{\partial X_j} \cdot \frac{X_j}{Y_R}$$

$$= \beta_0 \beta_j X_1^{\beta_1} X_2^{\beta_2} \cdots X_j^{\beta_j - 1} \cdots X_q^{\beta_q} \cdot \frac{X_j}{Y_R}$$

$$= \beta_j \cdot \frac{Y_R}{X_j} \cdot \frac{X_j}{Y_R}$$

$$= \beta_j$$

This means that, at the margin, if we change X_j by 1% and hold all other independent variables constant, Y_R will change by roughly β_j%. Elasticity therefore is constant and does not vary with the magnitude of Y_R or X_j.

NOTE 5.2

The common logarithm S of a number N is the power to which 10 must be raised to give N. Thus $\log N = S$ means that $10^S = N$. Similarly, the natural logarithm X of a number N is the power to which the constant $e (e = 2.71828 \ldots)$ must be raised to give N: $\ln N = X$ means that $e^X = N$. Operations with natural logarithms are exactly the same as with common logarithms; that is,

$$\ln(M \cdot N) = \ln M + \ln N$$

$$\ln(M/N) = \ln M - \ln N$$

$$\ln(M^p) = p \ln M$$

Some properties of natural logarithms are

$$\ln(e) = 1$$

$$\ln(1) = 0$$

$$\ln(0) = \text{not defined}$$

$$\ln(\text{negative number}) = \text{not defined}$$

A natural logarithm of a number N can readily be converted into a common logarithm of N and vise versa, if $10^S = e^X = N$, then

$$\log(10^S) = \log(e^X)$$

$$S \log 10 = X \log e$$

$$S = X \log e \qquad (\log e \approx .43429\ldots)$$

or

$$\ln(10^S) = \ln(e^X)$$

$$X = S \ln 10 \qquad (\ln 10 \approx 2.3025\ldots)$$

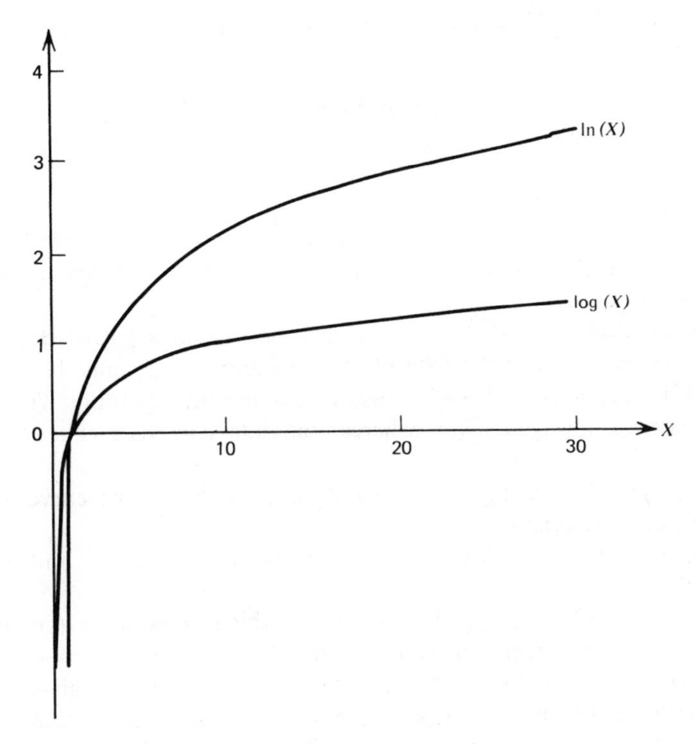

Figure A5.1

Figure A5.1 plots both types of logarithms against X. As examples, consider the transformations of Equations 1 and 3 in Table 5.4. First

$$Y = \exp(\beta_0 + \beta_1 X_1 + \cdots + \beta_q X_q) \cdot e^{\epsilon}$$

$$\ln Y = (\beta_0 + \beta_1 X_1 + \cdots + \beta_q X_q) \ln e + \epsilon \ln e$$

$$= \beta_0 + \beta_1 X_1 + \cdots + \beta_q X_q + \epsilon$$

Second,

$$Y = [1 + \beta_0 \exp(\beta_1 X_1 + \cdots + \beta_q X_q + \epsilon)]^{-1}$$

$$\frac{1}{Y} = 1 + \beta_0 \exp(\beta_1 X_1 + \cdots + \beta_q X_q + \epsilon)$$

$$\frac{1}{Y} - 1 = \beta_0 \exp(\beta_1 X_1 + \cdots + \beta_q X_q + \epsilon)$$

$$\ln\left(\frac{1}{Y} - 1\right) = \ln \beta_0 + (\beta_1 X_1 + \cdots + \beta_q X_q + \epsilon) \ln e$$

$$= \ln \beta_0 + \beta_1 X_1 + \cdots + \beta_q X_q + \epsilon$$

NOTE 5.3

(a) Consider first $Y_R = e^{\beta_0 + \beta_1 X_1}$. When $X_1 = 0$, $Y_R = e^{\beta_0}$, always a non-negative quantity.

Also, $dY_R/dX_1 = \beta_1 e^{\beta_1 X_1} \cdot e^{\beta_0}$, which is positive or negative, depending on the sign of β_1. To fit the data plotted in Figure 5.1 β_1 must be positive. Now, $d^2Y/dX_1^2 = \beta_1^2 e^{\beta_1 X_1} \cdot e^{\beta_0}$, always a nonnegative quantity. The shape of this curve is always convex, whereas the data suggests a concave curve (Figure A5.2):

(b) In $Y_R = \beta_0 X_1^{\beta_1}$, $Y_R = 0$, when X_1 is zero; hence the curve always goes through the origin.

Also, $dY_R/dX_1 = \beta_0 \beta_1 X_1^{\beta_1 - 1}$, which is positive if β_0, β_1, and X_1 are positive.

Now, $d^2Y_R/dX_1 = \beta_0 \beta_1(\beta_1 - 1)X_1^{\beta_1 - 2}$, which is negative (giving the required concave curve) if $\beta_1 < 1$, β_0, and X_1 are positive.

The "fit" obtained (Figure A5.3) is still not so good as that given by the polynomial in Figure 5.1; however, only one independent variable was used and the curve is more "predictable" than a polynomial.

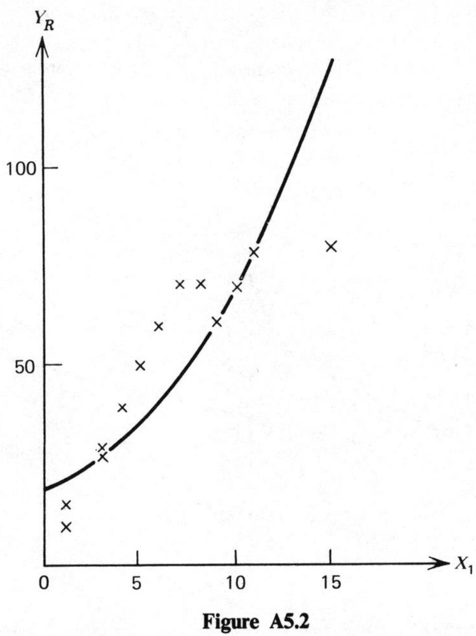

Figure A5.2

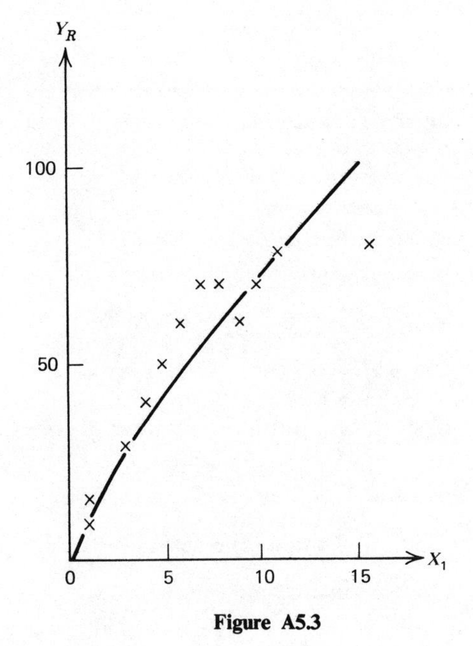

Figure A5.3

NOTE 5.4

If $Y_R = \beta_0 + \beta_1 X_1 + \delta_1 X_1 D_1$ were used, the relationship graphed in Figure A5.4 would have been fitted.

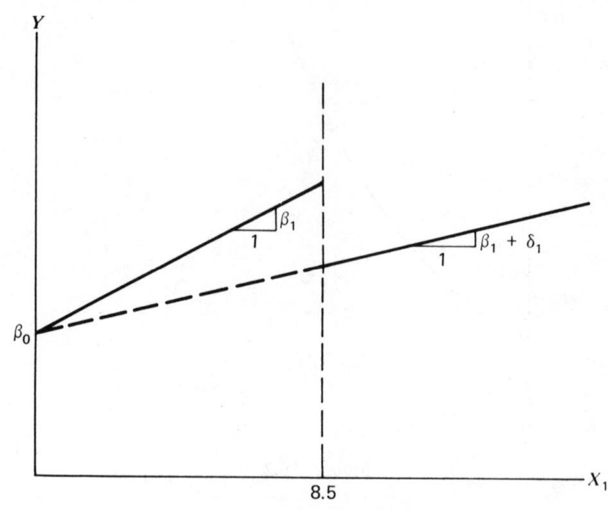

Figure A5.4

NOTE 6.1

Because $\epsilon_i = \rho\epsilon_{i-1} + \nu_i$, by substituting

$$\begin{aligned}\epsilon_i &= \rho(\rho\epsilon_{i-2} + \nu_{i-1}) + \nu_i \\ &= \rho^2\epsilon_{i-2} + \nu_i + \rho\nu_{i-1}\end{aligned}$$

Substituting again,

$$\begin{aligned}\epsilon_i &= \rho^2(\rho\epsilon_{i-3} + \nu_{i-2}) + \nu_i + \rho\nu_{i-1} \\ &= \rho^3\epsilon_{i-3} + \nu_i + \rho\nu_{i-1} + \rho^2\nu_{i-2}\end{aligned}$$

If we repeated this substitution an infinite number of times,

$$\epsilon_i = \rho^\infty\epsilon_{i-\infty} + \nu_i + \rho\nu_{i-1} + \rho^2\nu_{i-2} + \cdots$$

The first term on the right is 0 when $|\rho|$ is less than 1. Hence

$$\epsilon_i = \sum_{k=0}^{\infty} \rho^k \nu_{i-k}$$

(a) *Mean of* ϵ_i

$$E(\varepsilon_i) = E\left(\sum_{k=0}^{\infty} \rho^k \nu_{i-k}\right)$$

$$= \sum_{k=0}^{\infty} \rho^k E(\nu_{i-k})$$

$$= 0 \text{ since } E(\nu_{i-k}) \text{ is always } 0$$

(b) *Variance of* ϵ_i

$$\text{VAR}(\epsilon_i) = \text{VAR}\left(\sum_{k=0}^{\infty} \rho^k \nu_{i-k}\right)$$

Because the ν_{i-k} are statistically independent, according to Section 1.4, variance operation (3)

$$\text{VAR}(\epsilon_i) = \sum_{k=0}^{\infty} \text{VAR}(\rho^k \nu_{i-k})$$

$$= \sum_{k=0}^{\infty} (\rho^k)^2 \,\text{VAR}(\nu_{i-k})$$

$$= \sum_{k=0}^{\infty} \rho^{2k} \sigma_\nu^2$$

Using the formula for an infinite geometric series

$$\text{VAR}(\epsilon_i) = \sigma_\nu^2 \cdot \frac{1}{1-\rho^2}$$

(c) *Correlation Coefficient for* ϵ_i *and* ϵ_{i-1}. Using (1.16)

$$\text{COV}(\epsilon_i, \epsilon_{i-1}) = E[(\epsilon_i - 0)(\epsilon_{i-1} - 0)]$$

$$= E\left(\epsilon_i\left(\frac{\epsilon_i - \nu_i}{\rho}\right)\right)$$

$$= \frac{1}{\rho} E(\epsilon_i^2) - \frac{1}{\rho} E(\epsilon_i \nu_i)$$

$$= \frac{\sigma_\nu^2}{\rho}\left(\frac{1}{1-\rho^2}\right) - \frac{1}{\rho} E\left(\nu_i \sum_{k=0}^{\infty} \rho^k \nu_{i-k}\right)$$

$$= \frac{\sigma_\nu^2}{\rho}\left(\frac{1}{1-\rho^2}\right) - \frac{1}{\rho} \sum_{k=0}^{\infty} E(\nu_i \nu_{i-k}) \cdot \rho^k$$

Because $E(\nu_i \nu_{i-k}) = \sigma_\nu^2$ for $k = 0$ and zero otherwise (statistical independence),

$$\text{COV}(\varepsilon_i, \varepsilon_{i-1}) = \frac{\sigma_\nu^2}{\rho}\left(\frac{1}{1-\rho^2}\right) - \frac{\sigma_\nu^2}{\rho}$$

$$= \frac{\sigma_\nu^2 - \sigma_\nu^2 + \rho^2\sigma_\nu^2}{\rho(1-\rho^2)}$$

$$= \frac{\rho\sigma_\nu^2}{1-\rho^2}$$

The correlation coefficient is

$$\frac{\text{COV}(\epsilon_i, \epsilon_{i-1})}{\sqrt{\text{VAR}(\epsilon_i) \cdot \text{VAR}(\epsilon_{i-1})}} = \frac{\rho\sigma_\nu^2(1-\rho^2)}{\sigma_\nu^2(1-\rho^2)} = \rho$$

NOTE 6.2

The problem is to minimize

$$\sum_{i=2}^{n} (e_i - re_{i-1})^2$$

Setting the first derivative with respect to r to zero, we have

$$\sum 2(e_i - re_{i-1}) \cdot e_{i-1} = 0$$

$$\therefore \quad \sum e_i e_{i-1} - r \sum e_{i-1}^2 = 0$$

$$\therefore \quad r = \frac{\sum e_i \cdot e_{i-1}}{\sum e_{i-1}^2}$$

NOTE 6.3

$$D = \frac{\sum_{i=2}^{n} (e_i - e_{i-1})^2}{\sum_{i=1}^{n} e_i^2}$$

$$= \frac{\sum_{i=2}^{n} (e_i^2 - 2e_{i-1}e_i + e_{i-1}^2)}{\sum_{i=1}^{n} e_i^2}$$

$$= \frac{\sum_{i=2}^{n} e_i^2 + \sum_{i=2}^{n} e_{i-1}^2}{\sum_{i=1}^{n} e_i^2} - 2\frac{\sum_{i=2}^{n} e_{i-1}e_i}{\sum_{i=1}^{n} e_i^2}$$

$$= \frac{2\sum_{i=1}^{n} e_i^2 - e_1^2 - e_n^2}{\sum_{i=1}^{n} e_i^2} - 2\frac{\sum_{i=2}^{n} e_{i-1}e_i}{\sum_{i=2}^{n} e_{i-1}^2 + e_n^2}$$

If e_1^2 and e_n^2 are very small in relation to $\Sigma_{i=1}^{n} e_i^2$, then, using (6.9), $D \approx 2 - 2r$.

NOTE 7.1

To find least squares estimates $\hat{\mu}, \hat{\Upsilon}_1, \ldots, \hat{\Upsilon}_k$ for model (7.1) the sum of squared residuals

$$\sum_{j=1}^{k} \sum_{i=1}^{r_j} [Y_{ij} - (\hat{\mu} + \hat{\Upsilon}_j)]^2$$

must be minimized subject to the constraint $\Sigma_{j=1}^{k} \hat{\Upsilon}_j = 0$. These estimates will be identical to those found by a regression program [(7.4), (7.5)] in which the coefficients are found by minimizing the same sum of squared residuals under the same constraint.

This constrained minimization problem can be solved by the method of Lagrange multipliers. According to this method, we need merely find the extremal equations of L (if certain second-order conditions are met; they are met here) in which

$$L = \sum_{j=1}^{k} \sum_{i=1}^{r_j} [Y_{ij} - (\hat{\mu} + \hat{\Upsilon}_j)]^2 + \lambda\left(\sum_{j=1}^{k} \hat{\Upsilon}_j - 0\right)$$

where λ is a new variable. The extremal equations are

$$\frac{\partial L}{\partial \hat{\mu}} = -\sum_{j=1}^{k} \sum_{i=1}^{r_j} 2[Y_{ij} - (\hat{\mu} + \hat{\Upsilon}_j)] = 0$$

$$\frac{\partial L}{\partial \hat{\Upsilon}_j} = -\sum_{i=1}^{r_j} 2(Y_{ij} - (\hat{\mu} + \hat{\Upsilon}_j)) + \lambda = 0 \quad \text{for } j = 1, \ldots, k$$

$$\frac{\partial L}{\partial \lambda} = \sum_{j=1}^{k} \hat{\Upsilon}_j = 0$$

Note that if the sum of squared residuals were minimized without a constraint, the equation produced by $\partial L/\partial \hat{\mu}$ would be the same as the sum of the equations produced by $\partial L/\partial \hat{\Upsilon}$. We would actually have k equations in $k + 1$ unknowns and there would be an infinite number of solutions.

By adding the k equations derived from $\partial L/\partial \hat{\Upsilon}_j$ and then subtracting the first extremal equation it is seen that $k\lambda = 0$ or $\lambda = 0$. Now,

$$\sum_{i=1}^{r_j} [Y_{ij} - (\hat{\mu} + \hat{\Upsilon}_j)] = 0$$

$$\therefore \quad \sum_{i=1}^{r_j} (\hat{\mu} + \hat{\Upsilon}_j) = \sum_{i=1}^{r_j} Y_{ij}$$

$$\therefore \quad \hat{\mu} + \hat{\Upsilon}_j = \frac{\sum_{i=1}^{r_j} Y_{ij}}{r_j}$$

or

$$\hat{\Upsilon}_j = \frac{\sum_{i=1}^{r_j} Y_{ij}}{r_j} - \hat{\mu}$$

Also

$$\sum_{i=1}^{k} \hat{\Upsilon}_j = \sum_{j=1}^{k} \frac{\sum_{i=1}^{r_j} Y_{ij}}{r_j} - k\hat{\mu} = 0$$

$$\therefore \quad \hat{\mu} = \frac{1}{k} \cdot \sum_{j=1}^{k} \frac{\sum_{i=1}^{r_j} Y_{ij}}{r_j}$$

The expression for $\hat{\mu}$ states that the estimated overall mean is the average of the factor-level means. The expression for $\hat{\Upsilon}_j$ states that the estimated factor-level effect for factor j is the mean observation for the jth factor level minus the overall mean.

NOTE 7.2

The model

$$Y_{ij} = \mu' + \Upsilon_j' + \varepsilon_{ij}$$

where $\Sigma_{j=1}^{k} r_j \Upsilon_j' = 0$ is sometimes used. This note shows the correspondence between this model and (7.1). It also indicates the dummy variable coding necessary to obtain the estimates $\hat{\mu}'$, $\hat{\Upsilon}_1'$, . . . , $\hat{\Upsilon}_{k-1}'$ directly as regression coefficients. If all the r_j's are equal, the two models are identical.

Because both $\mu' + \Upsilon_j'$ and $\mu + \Upsilon_j$ indicate the expected value for Y when factor level j is in effect;

$$\mu' + \Upsilon_j' = \mu + \Upsilon_j = \mu_{\Upsilon_j}$$

Multiplying each side by r_j and adding all k equations, we have

$$r_1 \mu' + r_1 \Upsilon_1' = r_1 \mu_{\Upsilon_1}$$

$$\vdots$$

$$r_k \mu' + r_k \Upsilon_k' = r_k \mu_{\Upsilon_k}$$

Because $\sum r_j \Upsilon'_j = 0$

$$\mu' \sum r_j = \sum r_j \mu_{\Upsilon_j}$$

$$\mu' = \frac{\sum r_j \mu_{\Upsilon_j}}{\sum r_j}$$

The constant μ' can thus be interpreted as the weighted average of the expected values for Y. The weights are the numbers of observations for each factor level. Recall that μ is the unweighted average. Note that

$$\Upsilon'_j - \Upsilon_j = \mu - \mu'$$

It is evident that $\mu - \mu'$ does not depend on j; hence there is a constant difference between treatments estimated by the two methods. It also follows that

$$\Upsilon'_j - \Upsilon'_l = \Upsilon_j - \Upsilon_l$$

and differences among treatments are unaffected by the method chosen. Contrasts (Section 7.6) will also be the same.

Restated, the condition on the treatments is

$$r_1 \Upsilon'_1 + \cdots + r_k \Upsilon'_k = 0$$

$$\therefore \quad \Upsilon'_k = -\frac{r_1}{r_k} \Upsilon'_1 - \cdots - \frac{r_{k-1}}{r_k} \Upsilon'_{k-1}$$

The aim is to have the coefficient of X_j in the regression be Υ'_j. Let X_j be 1 if Y is observed under factor level j, where $j = 1, \ldots, k - 1$. For factor level k let

$$X_j = -\frac{r_j}{r_k}$$

It is seen (hopefully) that the required value of Υ'_k is produced. This coding is illustrated on the data matrix in Table 7.6, Example 7.4. The last five rows become

Y	X_1	X_2
21	− 8/5	− 5/5
32	− 8/5	− 5/5
17	− 8/5	− 5/5
24	− 8/5	− 5/5
15	− 8/5	− 5/5

Regression with the revised data yields

$$\hat{Y}_R = 31.556 + 9.444439X_1 - 5.355550X_2$$

The analysis of variance table is identical to Table 7.7. Therefore

$$\begin{aligned}\hat{\mu}' &= 31.56\\ \hat{\Upsilon}'_1 &= 9.44\\ \hat{\Upsilon}'_2 &= -\ 5.36\\ \hat{\Upsilon}'_3 &= -\ \tfrac{8}{5}(9.44) - \tfrac{5}{5}(-5.36)\\ &= -\ 9.76\end{aligned}$$

It can be verified that

$$\hat{\mu}' + \hat{\Upsilon}'_j = \hat{\mu} + \hat{\Upsilon}_j = \hat{\mu}_{\Upsilon_j}$$

by comparing these results with those of Example 7.4. It should be noted that μ' is estimated by the average of all observations, which is reasonable if we ponder the definition of μ'. The average observation in the example can be verified as 31.56. The estimation and hypothesis testing done in Examples 7.4 and 7.9 would give identical results if carried out with the model described above.

NOTE 7.3

If the interval

$$\hat{I}_{1-\alpha} = \left[\hat{L} - t_{\alpha/2}S_{\hat{L}},\ \hat{L} + t_{\alpha/2}S_{\hat{L}}\right]$$

at the $(1 - \alpha) \times 100\%$ confidence level does not contain L_0 and if

$$\begin{aligned}H_0 &: L = L_0\\ H_1 &: L \neq L_0\end{aligned}$$

H_0 will be rejected at the α level of significance.

Proof Consider Figure A7.1. Because $(\hat{L}_2 - L_0)/S_{\hat{L}} = t_{\alpha/2}$, $\hat{L}_2 = L_0 + t_{\alpha/2}S_{\hat{L}}$. Similarly, $\hat{L}_1 = L_0 - t_{\alpha/2}S_{\hat{L}}$. H_0 is rejected if $\hat{L} > \hat{L}_2$, giving $L_0 < \hat{L} - t_{\alpha/2}S_{\hat{L}}$, or if $\hat{L} < \hat{L}_1$, giving $L_0 > \hat{L} + t_{\alpha/2}S_{\hat{L}}$.

L_0 is not contained in $\hat{I}_{1-\alpha}$ if $L_0 > \hat{L} + t_{\alpha/2}S_{\hat{L}}$ or if $L_0 < \hat{L} - t_{\alpha/2}S_{\hat{L}}$. Therefore rejection of H_0 requires the same conditions as noninclusion of L_0 in $\hat{I}_{1-\alpha}$.

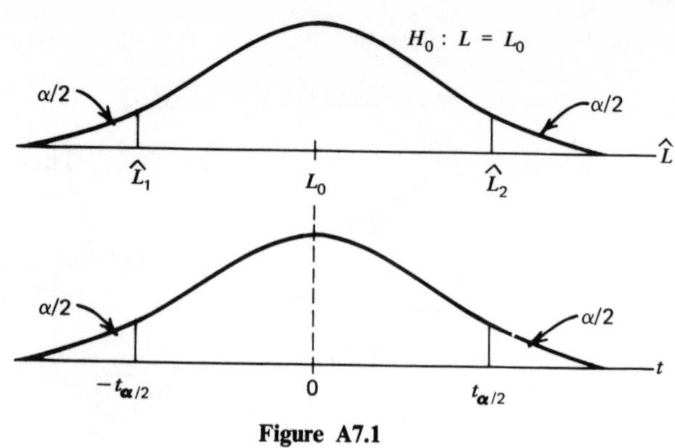

Figure A7.1

NOTE 8.1

Selected latin squares[1] (size axes):

3 × 3

A	B	C
B	C	A
C	A	B

4 × 4

1

A	B	C	D
B	A	D	C
C	D	B	A
D	C	A	B

2

A	B	C	D
B	C	D	A
C	D	A	B
D	A	B	C

3

A	B	C	D
B	D	A	C
C	A	D	B
D	C	B	A

4

A	B	C	D
B	A	D	C
C	D	A	B
D	C	B	A

5 × 5

A	B	C	D	E
B	A	E	C	D
C	D	A	E	B
D	E	B	A	C
E	C	D	B	A

6 × 6

A	B	C	D	E	F
B	F	D	C	A	E
C	D	E	F	B	A
D	A	F	E	C	B
E	C	A	B	F	D
F	E	B	A	D	C

7 × 7

A	B	C	D	E	F	G
B	C	D	E	F	G	A
C	D	E	F	G	A	B
D	E	F	G	A	B	C
E	F	G	A	B	C	D
F	G	A	B	C	D	E
G	A	B	C	D	E	F

8 × 8

A	B	C	D	E	F	G	H
B	C	D	E	F	G	H	A
C	D	E	F	G	H	A	B
D	E	F	G	H	A	B	C
E	F	G	H	A	B	C	D
F	G	H	A	B	C	D	E
G	H	A	B	C	D	E	F
H	A	B	C	D	E	F	G

9 × 9

A	B	C	D	E	F	G	H	I
B	C	D	E	F	G	H	I	A
C	D	E	F	G	H	I	A	B
D	E	F	G	H	I	A	B	C
E	F	G	H	I	A	B	C	D
F	G	H	I	A	B	C	D	E
G	H	I	A	B	C	D	E	F
H	I	A	B	C	D	E	F	G
I	A	B	C	D	E	F	G	H

[1] Adapted by permission from Plan 4.1, pp. 145–147, in William G. Cochran and Gertrude M. Cox, *Experimental Designs*. New York: Wiley, 1957.

10 × 10

A	B	C	D	E	F	G	H	I	J
B	C	D	E	F	G	H	I	J	A
C	D	E	F	G	H	I	J	A	B
D	E	F	G	H	I	J	A	B	C
E	F	G	H	I	J	A	B	C	D
F	G	H	I	J	A	B	C	D	E
G	H	I	J	A	B	C	D	E	F
H	I	J	A	B	C	D	E	F	G
I	J	A	B	C	D	E	F	G	H
J	A	B	C	D	E	F	G	H	I

11 × 11

A	B	C	D	E	F	G	H	I	J	K
B	C	D	E	F	G	H	I	J	K	A
C	D	E	F	G	H	I	J	K	A	B
D	E	F	G	H	I	J	K	A	B	C
E	F	G	H	I	J	K	A	B	C	D
F	G	H	I	J	K	A	B	C	D	E
G	H	I	J	K	A	B	C	D	E	F
H	I	J	K	A	B	C	D	E	F	G
I	J	K	A	B	C	D	E	F	G	H
J	K	A	B	C	D	E	F	G	H	I
K	A	B	C	D	E	F	G	H	I	J

12 × 12

A	B	C	D	E	F	G	H	I	J	K	L
B	C	D	E	F	G	H	I	J	K	L	A
C	D	E	F	G	H	I	J	K	L	A	B
D	E	F	G	H	I	J	K	L	A	B	C
E	F	G	H	I	J	K	L	A	B	C	D
F	G	H	I	J	K	L	A	B	C	D	E
G	H	I	J	K	L	A	B	C	D	E	F
H	I	J	K	L	A	B	C	D	E	F	G
I	J	K	L	A	B	C	D	E	F	G	H
J	K	L	A	B	C	D	E	F	G	H	I
K	L	A	B	C	D	E	F	G	H	I	J
L	A	B	C	D	E	F	G	H	I	J	K

Select a square of the required size (at random if $a = 4$). For $a \leqslant 4$ rearrange the rows at random; then rearrange the columns at random. For $a \geqslant 5$, after rearranging the rows and columns, rearrange the treatments at random as well.

This procedure does not select a latin square at random from all possible squares for $\alpha \geqslant 5$. However, the randomization is probably sufficient for practical purposes. See [8.3], 768–770, for examples.

NOTE 8.2

Suppose that the standard deviation σ of the error term is a function of μ, the mean of Y:

$$\sigma_Y = \phi(\mu)$$

Let $Z = f(Y)$ be a transformation intended to equalize variance.

By the first two terms of the Taylor expansion in the neighborhood of $Y = \mu$

$$Z \approx f(\mu) + f'(\mu)(Y - \mu)$$

$$\therefore \quad \mathrm{VAR}(Z) \approx [f'(\mu)]^2 \sigma_Y^2$$

$$\sigma_Z \approx f'(\mu)\sigma_Y$$

Because σ_Z is to be a constant

$$\sigma_Y f'(\mu) = C$$

where C is a constant, or

$$\phi(\mu)\, f'(\mu) = C$$

Given any assumed relationship $\phi(\mu)$, the differential equation is solved for the $f(\mu)$ [or $f(Y)$] required; for example, if $\sigma_Y^2 = k\mu$, where k is some constant, $f(Y) = \sqrt{Y}$ was the transformation given in Section 8.11. Checking, we have

$$\sqrt{k\mu} \cdot \frac{\mu^{-1/2}}{2} = \frac{\sqrt{k}}{2} \quad \text{which is a constant.}$$

appendix B

Matrices

B.1 MATRIX DEFINITIONS

A *matrix* is a rectangular array of numbers. Examples are

$$\begin{pmatrix} 3 & 2 \\ 6 & 1 \end{pmatrix} \quad \text{and} \quad \begin{pmatrix} 4 & 3 & 2 \\ 1 & 7 & 6 \end{pmatrix}$$

The entries in each array are called *elements*.

The *dimension* of a matrix specifies the number of its rows and columns. The first matrix has dimension 2×2, the second, 2×3. Matrices with the same dimensions are said to have the same *order*.

A matrix that has the same number of rows as columns is called a *square matrix*. Matrices of dimension 3×3, 12×12, 48×48, etc., are square matrices.

A square matrix with 1's in the main diagonal (upper left to lower right) and 0's everywhere else, is called an *identity matrix*; for example,

$$\begin{bmatrix} 1 & 0 & 0 \\ 0 & 1 & 0 \\ 0 & 0 & 1 \end{bmatrix} \quad \text{and} \quad \begin{bmatrix} 1 & 0 & 0 & 0 \\ 0 & 1 & 0 & 0 \\ 0 & 0 & 1 & 0 \\ 0 & 0 & 0 & 1 \end{bmatrix}$$

are identity matrices.

Matrices that have only one row or column are called *vectors*. A matrix containing only one row is a *row vector*. A matrix containing only one column is a *column vector* ((when the word vector is used alone, it usually refers to a column vector). Examples of row vectors are

$$(5 \quad 3 \quad 2) \quad \text{and} \quad (6 \quad 7 \quad 1 \quad 9 \quad 8 \quad 3)$$

Examples of column vectors are

$$\begin{bmatrix} 3 \\ 7 \\ 9 \end{bmatrix} \quad \text{and} \quad \begin{bmatrix} 1 \\ 5 \\ 7 \\ 10 \end{bmatrix}$$

B.2 MATRIX NOTATION

A single capital letter (some texts use boldface type) is used for a matrix, whereas its elements are represented by lower case letters subscripted first by row and then by column; for example,

$$A = \begin{pmatrix} a_{11} & a_{12} & a_{13} \\ a_{21} & a_{22} & a_{23} \end{pmatrix}$$

Two matrices of the same order are *equal* if the corresponding elements are equal. If $A = B$, where

$$A = \begin{pmatrix} a_{11} & a_{12} \\ a_{21} & a_{22} \end{pmatrix} \quad \text{and} \quad B = \begin{pmatrix} 3 & 7 \\ 4 & 0 \end{pmatrix}$$

it implies that $a_{11} = 3$, $a_{12} = 7$, $a_{21} = 4$, and $a_{22} = 0$.

The *transpose* of a matrix A is another matrix obtained by writing the columns of A as rows. A transpose of A is usually denoted as A' or A^T; for example:

$$A = \begin{pmatrix} 3 \\ 2 \end{pmatrix} \quad \text{and} \quad A' = (3 \quad 2)$$

$$B = \begin{bmatrix} 3 & 4 \\ 7 & 0 \\ 8 & 9 \end{bmatrix} \quad \text{and} \quad B' = \begin{pmatrix} 3 & 7 & 8 \\ 4 & 0 & 9 \end{pmatrix}$$

B.3 MATRIX OPERATIONS

Matrix addition (or *matrix subtraction*) is performed on two matrices of the same dimension by adding (or subtracting) the corresponding elements;

for example,

$$A = \begin{pmatrix} 3 & 2 \\ 7 & 1 \end{pmatrix} \quad \text{and} \quad B = \begin{pmatrix} 1 & 3 \\ 7 & -9 \end{pmatrix}$$

$$A + B = \begin{pmatrix} 3+1 & 2+3 \\ 7+7 & 1-9 \end{pmatrix}$$

$$A - B = \begin{pmatrix} 3-1 & 2-3 \\ 7-7 & 1+9 \end{pmatrix}$$

Multiplication by a scalar (an ordinary number or symbol for a number) is performed on a matrix by multiplying every element by that scalar; for example,

$$X = \begin{pmatrix} 7 & 6 \\ 6 & 6 \end{pmatrix} \quad \text{and} \quad 3X = \begin{pmatrix} 21 & 18 \\ 18 & 18 \end{pmatrix}$$

The *product* AB of two matrices A and B is defined only if the number of columns of A is equal to the number of rows of B. The operation is performed by taking dot products[1] of rows of A with columns of B. When a dot product is taken of row i of matrix A and column j in matrix B, the result is an element in the ith row and jth column of AB; for example,

$$A = \begin{pmatrix} 3 & 2 \\ 1 & 0 \end{pmatrix} \quad \text{and} \quad B = \begin{pmatrix} 1 \\ 2 \end{pmatrix}$$

then

$$AB = \begin{pmatrix} 3(1) + 2(2) \\ 1(1) + 0(2) \end{pmatrix} = \begin{pmatrix} 7 \\ 1 \end{pmatrix}$$

or

$$A = (5 \quad 6 \quad 7) \quad \text{and} \quad B = \begin{bmatrix} 1 & 3 \\ 1 & 4 \\ 0 & 8 \end{bmatrix}$$

then

$$AB = [5(1) + 6(1) + 7(0) \quad 5(3) + 6(4) + 7(8)]$$
$$= (11 \quad 95)$$

[1] A dot product of two vectors is best defined by an example: the dot product of $(a \quad b \quad c)$ and $(d \quad e \quad f)'$ is $ad + be + cf$.

Note that the product AB will be a matrix with the same number of rows as A and the same number of columns as B. Note also that although AB may exist, BA may not. Even if both AB and BA exist as they do for square matrices, the two products may not be the same (in AB we say that A is *postmultiplied* by B and that B is *premultiplied* by A); for example,

$$A = \begin{pmatrix} 2 & 5 \\ 0 & 1 \end{pmatrix} \quad \text{and} \quad B = \begin{pmatrix} 1 & 2 \\ 3 & 0 \end{pmatrix}$$

$$AB = \begin{pmatrix} 2(1) + 5(3) & 2(2) + 5(0) \\ 0(1) + 1(3) & 0(2) + 1(0) \end{pmatrix} = \begin{pmatrix} 17 & 4 \\ 3 & 0 \end{pmatrix}$$

$$BA = \begin{pmatrix} 1(2) + 2(0) & 1(5) + 2(1) \\ 3(2) + 0(0) & 3(5) + 0(1) \end{pmatrix} = \begin{pmatrix} 2 & 7 \\ 6 & 15 \end{pmatrix}$$

A square matrix premultiplied or postmultiplied by an identity matrix is unchanged; for example,

$$\begin{pmatrix} 1 & 0 \\ 0 & 1 \end{pmatrix}\begin{pmatrix} 3 & 2 \\ 7 & 6 \end{pmatrix} = \begin{pmatrix} 3 & 2 \\ 7 & 6 \end{pmatrix}$$

and

$$\begin{pmatrix} 3 & 2 \\ 7 & 6 \end{pmatrix}\begin{pmatrix} 1 & 0 \\ 0 & 1 \end{pmatrix} = \begin{pmatrix} 3 & 2 \\ 7 & 6 \end{pmatrix}$$

Matrix operations are useful in representing simultaneous linear equations; for example,

$$\begin{aligned} 3x_1 + 4x_2 + 7x_3 &= 7 \\ 5x_1 + 2x_2 \qquad\quad &= 2 \\ 3x_1 + 5x_2 + 8x_3 &= 0 \end{aligned}$$

can be represented as

$$\begin{bmatrix} 3 & 4 & 7 \\ 5 & 2 & 0 \\ 3 & 5 & 8 \end{bmatrix}\begin{bmatrix} x_1 \\ x_2 \\ x_3 \end{bmatrix} = \begin{bmatrix} 7 \\ 2 \\ 0 \end{bmatrix}$$

Performing the matrix multiplication,

$$\begin{bmatrix} 3x_1 + 4x_2 + 7x_3 \\ 5x_1 + 2x_2 + 0x_3 \\ 3x_1 + 5x_2 + 8x_3 \end{bmatrix} = \begin{bmatrix} 7 \\ 2 \\ 0 \end{bmatrix}$$

The definition of matrix equality now implies the original equations.

B.4 THE INVERSE OF A SQUARE MATRIX

The *inverse of a square matrix* A is another matrix, A^{-1}, of the same order such that when A is premultiplied or postmultiplied by A^{-1} an identity matrix results. As an example, if

$$A = \begin{pmatrix} 2 & 0 \\ 7 & -1 \end{pmatrix}$$

then

$$A^{-1} = \begin{pmatrix} \frac{1}{2} & 0 \\ \frac{7}{2} & -1 \end{pmatrix} \quad \text{is the inverse}.$$

Checking, we obtain

$$A^{-1}A = \begin{pmatrix} 1 & 0 \\ 0 & 1 \end{pmatrix}$$

and

$$AA^{-1} = \begin{pmatrix} 1 & 0 \\ 0 & 1 \end{pmatrix}$$

The inverse of a matrix does not always exist. A method of its calculation for a matrix of order 2 is given in Section B.6.

The inverse is useful for solving simultaneous linear equations. Consider the following:

$$\begin{bmatrix} a_{11} & \cdots & a_{1m} \\ \vdots & & \vdots \\ a_{n1} & & a_{nm} \end{bmatrix} \begin{bmatrix} x_1 \\ \vdots \\ x_n \end{bmatrix} = \begin{bmatrix} b_1 \\ \vdots \\ b_n \end{bmatrix}$$

or

$$AX = B$$

Premultiplying both sides by A^{-1} gives

$$A^{-1}AX = A^{-1}B$$

$$\therefore \quad X = A^{-1}B$$

To illustrate, suppose that

$$2x_1 \qquad = 4$$

$$7x_1 - x_2 = 7$$

or

$$\begin{pmatrix} 2 & 0 \\ 7 & -1 \end{pmatrix}\begin{pmatrix} x_1 \\ x_2 \end{pmatrix} = \begin{pmatrix} 4 \\ 7 \end{pmatrix}$$

Premultiplying both sides by the inverse $\begin{pmatrix} \frac{1}{2} & 0 \\ \frac{7}{2} & -1 \end{pmatrix}$ gives

$$\begin{pmatrix} x_1 \\ x_2 \end{pmatrix} = \begin{pmatrix} \frac{1}{2} & 0 \\ \frac{7}{2} & -1 \end{pmatrix}\begin{pmatrix} 4 \\ 7 \end{pmatrix}$$

$$= \begin{pmatrix} 2 \\ 7 \end{pmatrix}$$

That is, $x_1 = 2$, $x_2 = 7$.

B.5 THE DETERMINANT OF A MATRIX

The determinant (D) of a square matrix is a number calculated from the matrix; it may be negative, zero, or positive. For any 2×2 matrix

$$\begin{pmatrix} a & b \\ c & d \end{pmatrix}$$

$D = ad - bc$. For any 3×3 matrix

$$\begin{bmatrix} a & b & c \\ d & e & f \\ g & h & i \end{bmatrix}$$

$D = aei + bfg + chd - ahf - bdi - ceg$; for example, for

$$\begin{pmatrix} 3 & 2 \\ 7 & 1 \end{pmatrix}, \qquad D = 3 - 14 = -11$$

For

$$\begin{bmatrix} 7 & 3 & 1 \\ 0 & 0 & 7 \\ 1 & 5 & 1 \end{bmatrix}, \qquad D = 0 + 21 + 0 - 245 - 0 - 0 = -224$$

Determinants for matrices of any size may be calculated. The methods are not discussed here. If $D = 0$, the matrix is *singular*.

B.6 DETERMINANTS, INVERSES, AND REGRESSION

The inverse of a square matrix can be obtained by multiplying a second matrix, derived from the first, by $1/D$. A 2×2 matrix A where

$$A = \begin{pmatrix} a & b \\ c & d \end{pmatrix} \quad \text{gives} \quad A^{-1} = \frac{1}{D}\begin{pmatrix} d & -b \\ -c & a \end{pmatrix}$$

for example, if

$$A = \begin{pmatrix} 2 & 0 \\ 7 & -1 \end{pmatrix} \quad \text{then} \quad A^{-1} = \frac{1}{(-2)}\begin{pmatrix} -1 & 0 \\ -7 & 2 \end{pmatrix} = \begin{pmatrix} \frac{1}{2} & 0 \\ \frac{7}{2} & -1 \end{pmatrix}$$

If a matrix is singular, no inverse exists because $1/0$ is not defined.

In order to calculate regression coefficients a system of linear equations must be solved. Under certain circumstances a square matrix derived from the observations of independent variables is singular; hence the coefficients cannot be obtained (see Appendix A, Note 2.2).

appendix C

Tables

Table 1 Areas Under the Normal Curve

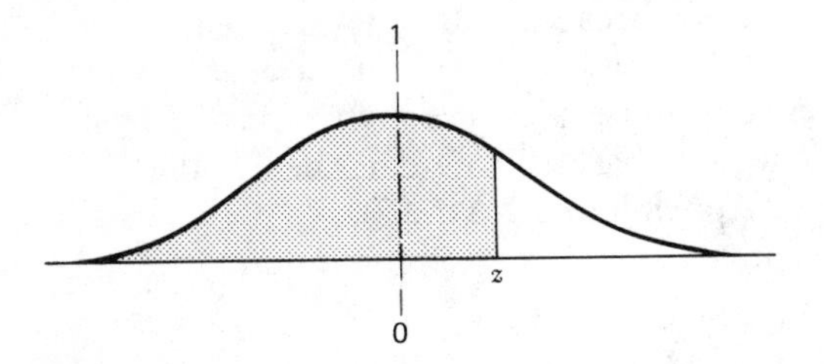

z	0.00	0.01	0.02	0.03	0.04	0.05	0.06	0.07	0.08	0.09
− 3.4	0.0003	0.0003	0.0003	0.0003	0.0003	0.0003	0.0003	0.0003	0.0003	0.0002
− 3.3	0.0005	0.0005	0.0005	0.0004	0.0004	0.0004	0.0004	0.0004	0.0004	0.0003
− 3.2	0.0007	0.0007	0.0006	0.0006	0.0006	0.0006	0.0006	0.0005	0.0005	0.0005
− 3.1	0.0010	0.0009	0.0009	0.0009	0.0008	0.0008	0.0008	0.0008	0.0007	0.0007
− 3.0	0.0013	0.0013	0.0013	0.0012	0.0012	0.0011	0.0011	0.0011	0.0010	0.0010
− 2.9	0.0019	0.0018	0.0017	0.0017	0.0016	0.0016	0.0015	0.0015	0.0014	0.0014
− 2.8	0.0026	0.0025	0.0024	0.0023	0.0023	0.0022	0.0021	0.0021	0.0020	0.0019
− 2.7	0.0035	0.0034	0.0033	0.0032	0.0031	0.0030	0.0029	0.0028	0.0027	0.0026
− 2.6	0.0047	0.0045	0.0044	0.0043	0.0041	0.0040	0.0039	0.0038	0.0037	0.0036
− 2.5	0.0062	0.0060	0.0059	0.0057	0.0055	0.0054	0.0052	0.0051	0.0049	0.0048
− 2.4	0.0082	0.0080	0.0078	0.0075	0.0073	0.0071	0.0069	0.0068	0.0066	0.0064
− 2.3	0.0107	0.0104	0.0102	0.0099	0.0096	0.0094	0.0091	0.0089	0.0087	0.0084
− 2.2	0.0139	0.0136	0.0132	0.0129	0.0125	0.0122	0.0119	0.0116	0.0113	0.0110
− 2.1	0.0179	0.0174	0.0170	0.0166	0.0162	0.0158	0.0154	0.0150	0.0146	0.0143
− 2.0	0.0228	0.0222	0.0217	0.0212	0.0207	0.0202	0.0197	0.0192	0.0188	0.0183
− 1.9	0.0287	0.0281	0.0274	0.0268	0.0262	0.0256	0.0250	0.0244	0.0239	0.0233
− 1.8	0.0359	0.0352	0.0344	0.0336	0.0329	0.0322	0.0314	0.0307	0.0301	0.0294
− 1.7	0.0446	0.0436	0.0427	0.0418	0.0409	0.0401	0.0392	0.0384	0.0375	0.0367
− 1.6	0.0548	0.0537	0.0526	0.0516	0.0505	0.0495	0.0485	0.0475	0.0465	0.0455
− 1.5	0.0668	0.0655	0.0643	0.0630	0.0618	0.0606	0.0594	0.0582	0.0571	0.0559

Table 1 (Continued)

z	0.00	0.01	0.02	0.03	0.04	0.05	0.06	0.07	0.08	0.09
− 1.4	0.0808	0.0793	0.0778	0.0764	0.0749	0.0735	0.0722	0.0708	0.0694	0.0681
− 1.3	0.0968	0.0951	0.0934	0.0918	0.0901	0.0885	0.0869	0.0853	0.0838	0.0823
− 1.2	0.1151	0.1131	0.1112	0.1093	0.1075	0.1056	0.1038	0.1020	0.1003	0.0985
− 1.1	0.1357	0.1335	0.1314	0.1292	0.1271	0.1251	0.1230	0.1210	0.1190	0.1170
− 1.0	0.1587	0.1562	0.1539	0.1515	0.1492	0.1469	0.1446	0.1423	0.1401	0.1379
− 0.9	0.1841	0.1814	0.1788	0.1762	0.1736	0.1711	0.1685	0.1660	0.1635	0.1611
− 0.8	0.2119	0.2090	0.2061	0.2033	0.2005	0.1977	0.1949	0.1922	0.1894	0.1867
− 0.7	0.2420	0.2389	0.2358	0.2327	0.2296	0.2266	0.2236	0.2206	0.2177	0.2148
− 0.6	0.2743	0.2709	0.2676	0.2643	0.2611	0.2578	0.2546	0.2514	0.2483	0.2451
− 0.5	0.3085	0.3050	0.3015	0.2981	0.2946	0.2912	0.2877	0.2843	0.2810	0.2776
− 0.4	0.3446	0.3409	0.3372	0.3336	0.3300	0.3264	0.3228	0.3192	0.3156	0.3121
− 0.3	0.3821	0.3783	0.3745	0.3707	0.3669	0.3632	0.3594	0.3557	0.3520	0.3483
− 0.2	0.4207	0.4168	0.4129	0.4090	0.4052	0.4013	0.3974	0.3936	0.3897	0.3859
− 0.1	0.4602	0.4562	0.4522	0.4483	0.4443	0.4404	0.4364	0.4325	0.4286	0.4247
− 0.0	0.5000	0.4960	0.4920	0.4880	0.4840	0.4801	0.4761	0.4721	0.4681	0.4641
0.0	0.5000	0.5040	0.5080	0.5120	0.5160	0.5199	0.5239	0.5279	0.5319	0.5359
0.1	0.5398	0.5438	0.5478	0.5517	0.5557	0.5596	0.5636	0.5675	0.5714	0.5753
0.2	0.5793	0.5832	0.5871	0.5910	0.5948	0.5987	0.6026	0.6064	0.6103	0.6141
0.3	0.6179	0.6217	0.6255	0.6293	0.6331	0.6368	0.6406	0.6443	0.6480	0.6517
0.4	0.6554	0.6591	0.6628	0.6664	0.6700	0.6736	0.6772	0.6808	0.6844	0.6879
0.5	0.6915	0.6950	0.6985	0.7019	0.7054	0.7088	0.7123	0.7157	0.7190	0.7224
0.6	0.7257	0.7291	0.7324	0.7357	0.7389	0.7422	0.7454	0.7486	0.7517	0.7549
0.7	0.7580	0.7611	0.7642	0.7673	0.7704	0.7734	0.7764	0.7794	0.7823	0.7852
0.8	0.7881	0.7910	0.7939	0.7967	0.7995	0.8023	0.8051	0.8078	0.8106	0.8133
0.9	0.8159	0.8186	0.8212	0.8238	0.8264	0.8289	0.8315	0.8340	0.8365	0.8389
1.0	0.8413	0.8438	0.8461	0.8485	0.8508	0.8531	0.8554	0.8577	0.8599	0.8621
1.1	0.8643	0.8665	0.8686	0.8708	0.8729	0.8749	0.8770	0.8790	0.8810	0.8830
1.2	0.8849	0.8869	0.8888	0.8907	0.8925	0.8944	0.8962	0.8980	0.8997	0.9015
1.3	0.9032	0.9049	0.9066	0.9082	0.9099	0.9115	0.9131	0.9147	0.9162	0.9177
1.4	0.9192	0.9207	0.9222	0.9236	0.9251	0.9265	0.9278	0.9292	0.9306	0.9319
1.5	0.9332	0.9345	0.9357	0.9370	0.9382	0.9394	0.9406	0.9418	0.9429	0.9441
1.6	0.9452	0.9463	0.9474	0.9484	0.9495	0.9505	0.9515	0.9525	0.9535	0.9545
1.7	0.9554	0.9564	0.9573	0.9582	0.9591	0.9599	0.9608	0.9616	0.9625	0.9633
1.8	0.9641	0.9649	0.9656	0.9664	0.9671	0.9678	0.9686	0.9693	0.9699	0.9706
1.9	0.9713	0.9719	0.9726	0.9732	0.9738	0.9744	0.9750	0.9756	0.9761	0.9767
2.0	0.9772	0.9778	0.9783	0.9788	0.9793	0.9798	0.9803	0.9808	0.9812	0.9817
2.1	0.9821	0.9826	0.9830	0.9834	0.9838	0.9842	0.9846	0.9850	0.9854	0.9857
2.2	0.9861	0.9864	0.9868	0.9871	0.9875	0.9878	0.9881	0.9884	0.9887	0.9890
2.3	0.9893	0.9896	0.9898	0.9901	0.9904	0.9906	0.9909	0.9911	0.9913	0.9916
2.4	0.9918	0.9920	0.9922	0.9925	0.9927	0.9929	0.9931	0.9932	0.9934	0.9936

Table 1 (Continued)

z	0.00	0.01	0.02	0.03	0.04	0.05	0.06	0.07	0.08	0.09
2.5	0.9938	0.9940	0.9941	0.9943	0.9945	0.9946	0.9948	0.9949	0.9951	0.9952
2.6	0.9953	0.9955	0.9956	0.9957	0.9959	0.9960	0.9961	0.9962	0.9963	0.9964
2.7	0.9965	0.9966	0.9967	0.9968	0.9969	0.9970	0.9971	0.9972	0.9973	0.9974
2.8	0.9974	0.9975	0.9976	0.9977	0.9977	0.9978	0.9979	0.9979	0.9980	0.9981
2.9	0.9981	0.9982	0.9982	0.9983	0.9984	0.9984	0.9985	0.9985	0.9986	0.9986
3.0	0.9987	0.9987	0.9987	0.9988	0.9988	0.9989	0.9989	0.9989	0.9990	0.9990
3.1	0.9990	0.9991	0.9991	0.9991	0.9992	0.9992	0.9992	0.9992	0.9993	0.9993
3.2	0.9993	0.9993	0.9994	0.9994	0.9994	0.9994	0.9994	0.9995	0.9995	0.9995
3.3	0.9995	0.9995	0.9995	0.9996	0.9996	0.9996	0.9996	0.9996	0.9996	0.9997
3.4	0.9997	0.9997	0.9997	0.9997	0.9997	0.9997	0.9997	0.9997	0.9997	0.9998

Taken by permission from Ronald E. Walpole, *Introduction to Statistics*, Table A4. New York, The Macmillan Company, 1968.

Table 2 Student's t-Distribution

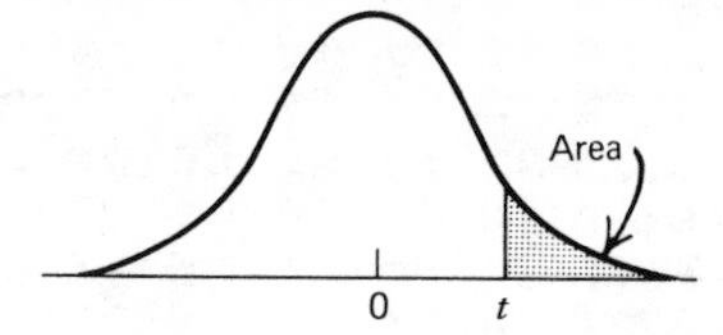

Number of degrees of freedom	Value of area					
	0.25	0.10	0.05	0.025	0.010	0.005
1	1.000	3.078	6.314	12.706	31.821	63.657
2	0.816	1.886	2.920	4.303	6.965	9.925
3	0.765	1.638	2.353	3.182	4.541	5.841
4	0.741	1.533	2.132	2.776	3.747	4.604
5	0.727	1.476	2.015	2.571	3.365	4.032
6	0.718	1.440	1.943	2.447	3.143	3.707
7	0.711	1.415	1.895	2.365	2.998	3.499
8	0.706	1.397	1.860	2.306	2.896	3.355
9	0.703	1.383	1.833	2.262	2.821	3.250
10	0.700	1.372	1.812	2.228	2.764	3.169
11	0.697	1.363	1.796	2.201	2.718	3.106
12	0.695	1.356	1.782	2.179	2.681	3.055
13	0.694	1.350	1.771	2.160	2.650	3.012
14	0.692	1.345	1.761	2.145	2.624	2.977
15	0.691	1.341	1.753	2.131	2.602	2.947
16	0.690	1.337	1.746	2.120	2.583	2.921
17	0.689	1.333	1.740	2.110	2.567	2.898
18	0.688	1.330	1.734	2.101	2.552	2.878
19	0.688	1.328	1.729	2.093	2.539	2.861
20	0.687	1.325	1.725	2.086	2.528	2.845
21	0.686	1.323	1.721	2.080	2.518	2.831
22	0.686	1.321	1.717	2.074	2.508	2.819
23	0.685	1.319	1.714	2.069	2.500	2.807
24	0.685	1.318	1.711	2.064	2.492	2.797
25	0.684	1.316	1.708	2.060	2.485	2.787
26	0.684	1.315	1.706	2.056	2.479	2.779
27	0.684	1.314	1.703	2.052	2.473	2.771
28	0.683	1.313	1.701	2.048	2.467	2.763
29	0.683	1.311	1.699	2.045	2.462	2.756
30	0.683	1.310	1.697	2.042	2.457	2.750
∞	0.674	1.282	1.645	1.960	2.326	2.576

Source: by permission from R. A. Fisher, *Statistical Methods for Research Workers*, 14th ed., Table IV. New York: Hafner Press, 1972.

Table 3 *F* Distribution. Entries in the table are values of *F* for which area in upper tail is 0.05 (roman type) or 0.01 (boldface type).

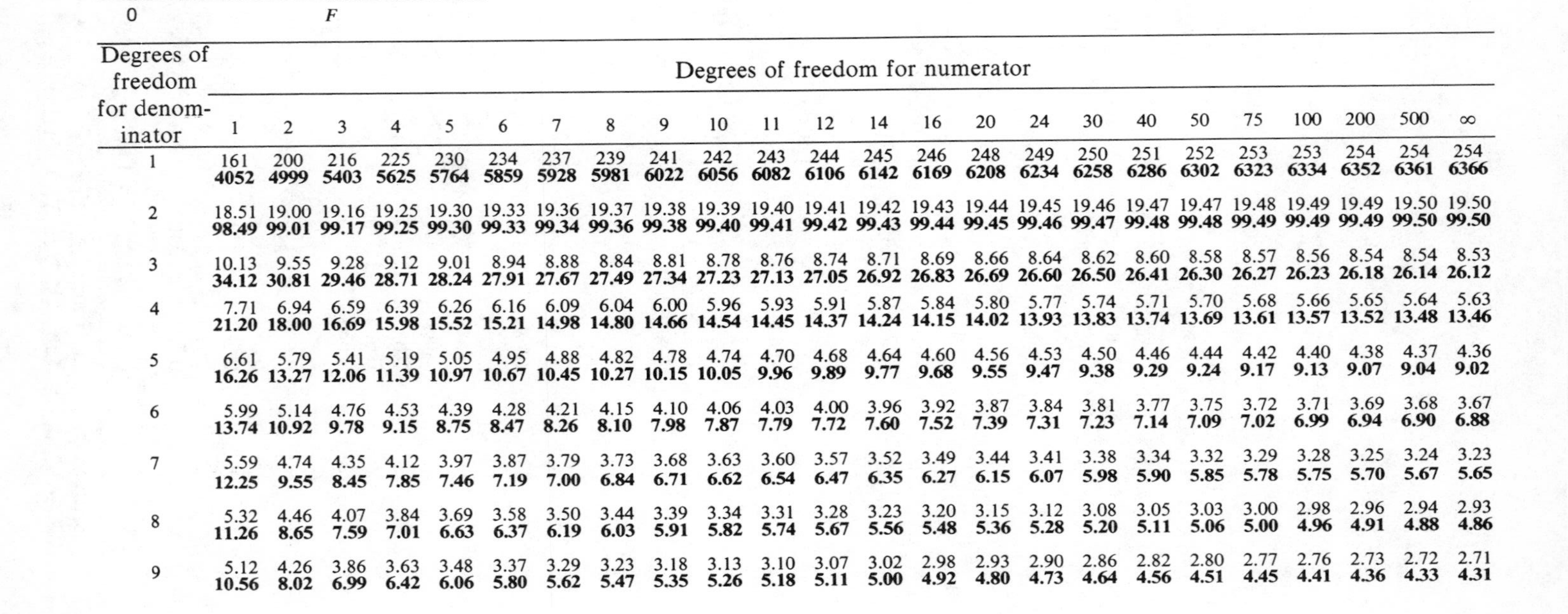

Degrees of freedom for denominator	Degrees of freedom for numerator																							
	1	2	3	4	5	6	7	8	9	10	11	12	14	16	20	24	30	40	50	75	100	200	500	∞
1	161 **4052**	200 **4999**	216 **5403**	225 **5625**	230 **5764**	234 **5859**	237 **5928**	239 **5981**	241 **6022**	242 **6056**	243 **6082**	244 **6106**	245 **6142**	246 **6169**	248 **6208**	249 **6234**	250 **6258**	251 **6286**	252 **6302**	253 **6323**	253 **6334**	254 **6352**	254 **6361**	254 **6366**
2	18.51 **98.49**	19.00 **99.01**	19.16 **99.17**	19.25 **99.25**	19.30 **99.30**	19.33 **99.33**	19.36 **99.34**	19.37 **99.36**	19.38 **99.38**	19.39 **99.40**	19.40 **99.41**	19.41 **99.42**	19.42 **99.43**	19.43 **99.44**	19.44 **99.45**	19.45 **99.46**	19.46 **99.47**	19.47 **99.48**	19.47 **99.48**	19.48 **99.49**	19.49 **99.49**	19.49 **99.49**	19.50 **99.50**	19.50 **99.50**
3	10.13 **34.12**	9.55 **30.81**	9.28 **29.46**	9.12 **28.71**	9.01 **28.24**	8.94 **27.91**	8.88 **27.67**	8.84 **27.49**	8.81 **27.34**	8.78 **27.23**	8.76 **27.13**	8.74 **27.05**	8.71 **26.92**	8.69 **26.83**	8.66 **26.69**	8.64 **26.60**	8.62 **26.50**	8.60 **26.41**	8.58 **26.30**	8.57 **26.27**	8.56 **26.23**	8.54 **26.18**	8.54 **26.14**	8.53 **26.12**
4	7.71 **21.20**	6.94 **18.00**	6.59 **16.69**	6.39 **15.98**	6.26 **15.52**	6.16 **15.21**	6.09 **14.98**	6.04 **14.80**	6.00 **14.66**	5.96 **14.54**	5.93 **14.45**	5.91 **14.37**	5.87 **14.24**	5.84 **14.15**	5.80 **14.02**	5.77 **13.93**	5.74 **13.83**	5.71 **13.74**	5.70 **13.69**	5.68 **13.61**	5.66 **13.57**	5.65 **13.52**	5.64 **13.48**	5.63 **13.46**
5	6.61 **16.26**	5.79 **13.27**	5.41 **12.06**	5.19 **11.39**	5.05 **10.97**	4.95 **10.67**	4.88 **10.45**	4.82 **10.27**	4.78 **10.15**	4.74 **10.05**	4.70 **9.96**	4.68 **9.89**	4.64 **9.77**	4.60 **9.68**	4.56 **9.55**	4.53 **9.47**	4.50 **9.38**	4.46 **9.29**	4.44 **9.24**	4.42 **9.17**	4.40 **9.13**	4.38 **9.07**	4.37 **9.04**	4.36 **9.02**
6	5.99 **13.74**	5.14 **10.92**	4.76 **9.78**	4.53 **9.15**	4.39 **8.75**	4.28 **8.47**	4.21 **8.26**	4.15 **8.10**	4.10 **7.98**	4.06 **7.87**	4.03 **7.79**	4.00 **7.72**	3.96 **7.60**	3.92 **7.52**	3.87 **7.39**	3.84 **7.31**	3.81 **7.23**	3.77 **7.14**	3.75 **7.09**	3.72 **7.02**	3.71 **6.99**	3.69 **6.94**	3.68 **6.90**	3.67 **6.88**
7	5.59 **12.25**	4.74 **9.55**	4.35 **8.45**	4.12 **7.85**	3.97 **7.46**	3.87 **7.19**	3.79 **7.00**	3.73 **6.84**	3.68 **6.71**	3.63 **6.62**	3.60 **6.54**	3.57 **6.47**	3.52 **6.35**	3.49 **6.27**	3.44 **6.15**	3.41 **6.07**	3.38 **5.98**	3.34 **5.90**	3.32 **5.85**	3.29 **5.78**	3.28 **5.75**	3.25 **5.70**	3.24 **5.67**	3.23 **5.65**
8	5.32 **11.26**	4.46 **8.65**	4.07 **7.59**	3.84 **7.01**	3.69 **6.63**	3.58 **6.37**	3.50 **6.19**	3.44 **6.03**	3.39 **5.91**	3.34 **5.82**	3.31 **5.74**	3.28 **5.67**	3.23 **5.56**	3.20 **5.48**	3.15 **5.36**	3.12 **5.28**	3.08 **5.20**	3.05 **5.11**	3.03 **5.06**	3.00 **5.00**	2.98 **4.96**	2.96 **4.91**	2.94 **4.88**	2.93 **4.86**
9	5.12 **10.56**	4.26 **8.02**	3.86 **6.99**	3.63 **6.42**	3.48 **6.06**	3.37 **5.80**	3.29 **5.62**	3.23 **5.47**	3.18 **5.35**	3.13 **5.26**	3.10 **5.18**	3.07 **5.11**	3.02 **5.00**	2.98 **4.92**	2.93 **4.80**	2.90 **4.73**	2.86 **4.64**	2.82 **4.56**	2.80 **4.51**	2.77 **4.45**	2.76 **4.41**	2.73 **4.36**	2.72 **4.33**	2.71 **4.31**

10	4.96	4.10	3.71	3.48	3.33	3.22	3.14	3.07	3.02	2.97	2.94	2.91	2.86	2.82	2.77	2.74	2.70	2.67	2.64	2.61	2.59	2.56	2.55	2.54
	10.04	**7.56**	**6.55**	**5.99**	**5.64**	**5.39**	**5.21**	**5.06**	**4.95**	**4.85**	**4.78**	**4.71**	**4.60**	**4.52**	**4.41**	**4.33**	**4.25**	**4.17**	**4.12**	**4.05**	**4.01**	**3.96**	**3.93**	**3.91**
11	4.84	3.98	3.59	3.36	3.20	3.09	3.01	2.95	2.90	2.86	2.82	2.79	2.74	2.70	2.65	2.61	2.57	2.53	2.50	2.47	2.45	2.42	2.41	2.40
	9.65	**7.20**	**6.22**	**5.67**	**5.32**	**5.07**	**4.88**	**4.74**	**4.63**	**4.54**	**4.46**	**4.40**	**4.29**	**4.21**	**4.10**	**4.02**	**3.94**	**3.86**	**3.80**	**3.74**	**3.70**	**3.66**	**3.62**	**3.60**
12	4.75	3.88	3.49	3.26	3.11	3.00	2.92	2.85	2.80	2.76	2.72	2.69	2.64	2.60	2.54	2.50	2.46	2.42	2.40	2.36	2.35	2.32	2.31	2.30
	9.33	**6.93**	**5.95**	**5.41**	**5.06**	**4.82**	**4.65**	**4.50**	**4.39**	**4.30**	**4.22**	**4.16**	**4.05**	**3.98**	**3.86**	**3.78**	**3.70**	**3.61**	**3.56**	**3.49**	**3.46**	**3.41**	**3.38**	**3.36**
13	4.67	3.80	3.41	3.18	3.02	2.92	2.81	2.77	2.72	2.67	2.63	2.60	2.55	2.51	2.46	2.42	2.38	2.34	2.32	2.28	2.26	2.24	2.22	2.21
	9.07	**6.70**	**5.74**	**5.20**	**4.86**	**4.62**	**4.44**	**4.30**	**4.19**	**4.10**	**4.02**	**3.96**	**3.85**	**3.78**	**3.67**	**3.59**	**3.51**	**3.42**	**3.37**	**3.30**	**3.27**	**3.21**	**3.18**	**3.16**
14	4.60	3.74	3.34	3.11	2.96	2.85	2.77	2.70	2.65	2.60	2.56	2.53	2.48	2.44	2.39	2.35	2.31	2.27	2.24	2.21	2.19	2.16	2.14	2.13
	8.86	**6.51**	**5.56**	**5.03**	**4.69**	**4.46**	**4.28**	**4.14**	**4.03**	**3.94**	**3.86**	**3.80**	**3.70**	**3.62**	**3.51**	**3.43**	**3.34**	**3.26**	**3.21**	**3.14**	**3.11**	**3.06**	**3.02**	**3.00**
15	4.54	3.68	3.29	3.06	2.90	2.79	2.70	2.64	2.59	2.55	2.51	2.48	2.43	2.39	2.33	2.29	2.25	2.21	2.18	2.15	2.12	2.10	2.08	2.07
	8.68	**6.36**	**5.42**	**4.89**	**4.56**	**4.33**	**4.14**	**4.00**	**3.89**	**3.80**	**3.73**	**3.67**	**3.56**	**3.48**	**3.36**	**3.29**	**3.20**	**3.12**	**3.07**	**3.00**	**2.97**	**2.92**	**2.89**	**2.87**
16	4.49	3.63	3.24	3.01	2.85	2.74	2.66	2.59	2.54	2.49	2.45	2.42	2.37	2.33	2.28	2.24	2.20	2.16	2.13	2.09	2.07	2.04	2.02	2.01
	8.53	**6.23**	**5.29**	**4.77**	**4.44**	**4.20**	**4.03**	**3.89**	**3.78**	**3.69**	**3.61**	**3.55**	**3.45**	**3.37**	**3.25**	**3.18**	**3.10**	**3.01**	**2.96**	**2.89**	**2.86**	**2.80**	**2.77**	**2.75**
17	4.45	3.59	3.20	2.96	2.81	2.70	2.62	2.55	2.50	2.45	2.41	2.38	2.33	2.29	2.23	2.19	2.15	2.11	2.08	2.04	2.02	1.99	1.97	1.96
	8.40	**6.11**	**5.18**	**4.67**	**4.34**	**4.10**	**3.93**	**3.79**	**3.68**	**3.59**	**3.52**	**3.45**	**3.35**	**3.27**	**3.16**	**3.08**	**3.00**	**2.92**	**2.86**	**2.79**	**2.76**	**2.70**	**2.67**	**2.65**
18	4.41	3.55	3.16	2.93	2.77	2.66	2.58	2.51	2.46	2.41	2.37	2.34	2.29	2.25	2.19	2.15	2.11	2.07	2.04	2.00	1.98	1.95	1.93	1.92
	8.28	**6.01**	**5.09**	**4.58**	**4.25**	**4.01**	**3.85**	**3.71**	**3.60**	**3.51**	**3.44**	**3.37**	**3.27**	**3.19**	**3.07**	**3.00**	**2.91**	**2.83**	**2.78**	**2.71**	**2.68**	**2.62**	**2.59**	**2.57**
19	4.38	3.52	3.13	2.90	2.74	2.63	2.55	2.48	2.43	2.38	2.34	2.31	2.20	2.21	2.15	2.11	2.07	2.02	2.00	1.96	1.94	1.91	1.90	1.88
	8.18	**6.93**	**5.01**	**4.50**	**4.17**	**3.94**	**3.77**	**3.63**	**3.52**	**3.43**	**3.36**	**3.30**	**3.19**	**3.12**	**3.00**	**2.92**	**2.84**	**2.76**	**2.70**	**2.63**	**2.60**	**2.54**	**2.51**	**2.49**
20	4.35	3.49	3.10	2.87	2.71	2.60	2.52	2.45	2.40	2.35	2.31	2.28	2.23	2.18	2.12	2.08	2.04	1.99	1.96	1.92	1.90	1.87	1.85	1.84
	8.10	**5.85**	**4.94**	**4.43**	**4.10**	**3.87**	**3.71**	**3.56**	**3.45**	**3.37**	**3.30**	**3.23**	**3.13**	**3.05**	**2.94**	**2.86**	**2.77**	**2.69**	**2.63**	**2.56**	**2.53**	**2.47**	**2.44**	**2.42**
21	4.32	3.47	3.07	2.84	2.68	2.57	2.49	2.42	2.37	2.32	2.28	2.25	2.20	2.15	2.09	2.05	2.00	1.96	1.93	1.89	1.87	1.84	1.82	1.81
	8.02	**5.78**	**4.87**	**4.37**	**4.04**	**3.81**	**3.65**	**3.51**	**3.40**	**3.31**	**3.24**	**3.17**	**3.07**	**2.99**	**2.88**	**2.80**	**2.72**	**2.63**	**2.58**	**2.51**	**2.47**	**2.42**	**2.38**	**2.36**
22	4.30	3.44	3.05	2.82	2.66	2.55	2.47	2.40	2.35	2.30	2.26	2.23	2.18	2.13	2.07	2.03	1.98	1.93	1.91	1.87	1.84	1.81	1.80	1.78
	7.94	**5.72**	**4.82**	**4.31**	**3.99**	**3.76**	**3.59**	**3.45**	**3.35**	**3.26**	**3.18**	**3.12**	**3.02**	**2.94**	**2.83**	**2.75**	**2.67**	**2.58**	**2.53**	**2.46**	**2.42**	**2.37**	**2.33**	**2.31**
23	4.28	3.42	3.03	2.80	2.64	2.53	2.45	2.38	2.32	2.28	2.24	2.20	2.14	2.10	2.04	2.00	1.96	1.91	1.88	1.84	1.82	1.79	1.77	1.76
	7.88	**5.66**	**4.76**	**4.26**	**3.94**	**3.71**	**3.54**	**3.41**	**3.30**	**3.21**	**3.14**	**3.07**	**2.97**	**2.89**	**2.78**	**2.70**	**2.62**	**2.53**	**2.48**	**2.41**	**2.37**	**2.32**	**2.28**	**2.26**
24	4.26	3.40	3.01	2.78	2.62	2.51	2.43	2.36	2.30	2.26	2.22	2.18	2.13	2.09	2.02	1.98	1.94	1.89	1.86	1.82	1.80	1.76	1.74	1.73
	7.82	**5.61**	**4.72**	**4.22**	**3.90**	**3.67**	**3.50**	**3.36**	**3.25**	**3.17**	**3.09**	**3.03**	**2.93**	**2.85**	**2.74**	**2.66**	**2.58**	**2.49**	**2.44**	**2.36**	**2.33**	**2.27**	**2.23**	**2.21**
25	4.24	3.38	2.99	2.76	2.60	2.49	2.41	2.34	2.28	2.24	2.20	2.16	2.11	2.06	2.00	1.96	1.92	1.87	1.84	1.80	1.77	1.74	1.72	1.71
	7.77	**5.57**	**4.68**	**4.18**	**3.86**	**3.63**	**3.46**	**3.32**	**3.21**	**3.13**	**3.05**	**2.99**	**2.89**	**2.81**	**2.70**	**2.62**	**2.54**	**2.45**	**2.40**	**2.32**	**2.29**	**2.23**	**2.19**	**2.17**
26	4.22	3.37	2.89	2.74	2.59	2.47	2.39	2.32	2.27	2.22	2.18	2.15	2.10	2.05	1.99	1.95	1.90	1.85	1.82	1.78	1.76	1.72	1.70	1.69
	7.72	**5.53**	**4.64**	**4.14**	**3.82**	**3.59**	**3.42**	**3.29**	**3.17**	**3.09**	**3.02**	**2.96**	**2.86**	**2.77**	**2.66**	**2.58**	**2.50**	**2.41**	**2.36**	**2.28**	**2.25**	**2.19**	**2.15**	**2.13**

Table 3 (continued)

Degrees of freedom for denominator	Degrees of freedom for numerator																							
	1	2	3	4	5	6	7	8	9	10	11	12	14	16	20	24	30	40	50	75	100	200	500	∞
27	4.21 **7.68**	3.35 **5.49**	2.96 **4.60**	2.73 **4.11**	2.57 **3.79**	2.46 **3.56**	2.37 **3.39**	2.30 **3.26**	2.25 **3.14**	2.20 **3.06**	2.16 **2.08**	2.13 **2.93**	2.08 **2.83**	2.03 **2.74**	1.97 **2.63**	1.93 **2.55**	1.88 **2.47**	1.84 **2.38**	1.80 **2.33**	1.76 **2.25**	1.74 **2.21**	1.71 **2.16**	1.68 **2.12**	1.67 **2.10**
28	4.20 **7.64**	3.34 **5.45**	2.95 **4.57**	2.71 **4.07**	2.56 **3.76**	2.44 **3.53**	2.36 **3.36**	2.29 **3.23**	3.24 **3.11**	2.19 **3.03**	2.15 **2.95**	2.12 **2.90**	2.06 **2.80**	2.02 **2.71**	1.96 **2.60**	1.91 **2.52**	1.87 **2.44**	1.81 **2.35**	1.78 **2.30**	1.75 **2.22**	1.72 **2.18**	1.69 **2.13**	1.67 **2.09**	1.65 **2.06**
29	4.18 **7.60**	3.33 **5.52**	2.93 **4.54**	2.70 **4.04**	2.54 **3.73**	2.43 **3.50**	2.35 **3.33**	2.28 **3.20**	2.22 **3.08**	2.18 **3.00**	2.14 **2.92**	2.10 **2.87**	2.05 **2.77**	2.00 **2.68**	1.94 **2.57**	1.90 **2.49**	1.85 **2.41**	1.80 **2.32**	1.77 **2.27**	1.73 **2.19**	1.71 **2.15**	1.68 **2.10**	1.65 **2.06**	1.64 **2.03**
30	4.17 **7.56**	3.32 **5.39**	2.92 **4.51**	2.69 **4.02**	2.53 **3.70**	2.42 **3.47**	2.34 **3.30**	2.27 **3.17**	2.21 **3.06**	2.16 **2.98**	2.12 **2.00**	2.09 **2.84**	2.04 **2.74**	1.99 **2.66**	1.93 **2.55**	1.89 **2.47**	1.84 **2.38**	1.79 **2.29**	1.76 **2.24**	1.72 **2.16**	1.69 **2.13**	1.66 **2.07**	1.64 **2.03**	1.62 **2.01**
32	4.15 **7.50**	3.30 **5.34**	2.90 **4.46**	2.67 **3.97**	2.51 **3.66**	2.40 **3.42**	2.32 **3.25**	2.25 **3.12**	2.19 **3.01**	2.14 **2.94**	2.10 **2.86**	2.07 **2.80**	2.02 **2.70**	1.97 **2.62**	1.91 **2.51**	1.86 **2.42**	1.82 **2.34**	1.76 **2.25**	1.74 **2.20**	1.69 **2.12**	1.67 **2.08**	1.64 **2.02**	1.61 **1.98**	1.59 **1.96**
34	4.13 **7.44**	3.28 **5.29**	2.88 **4.42**	2.65 **3.93**	2.49 **3.61**	2.38 **3.38**	2.30 **3.21**	2.23 **3.08**	2.17 **2.97**	2.12 **2.89**	2.08 **2.82**	2.05 **2.76**	2.00 **2.66**	1.95 **2.58**	1.89 **2.47**	1.84 **2.38**	1.80 **2.30**	1.74 **2.21**	1.71 **2.15**	1.67 **2.08**	1.64 **2.04**	1.61 **1.98**	1.59 **1.94**	1.57 **1.91**
36	4.11 **7.39**	3.26 **5.25**	2.86 **4.38**	2.63 **3.89**	2.48 **3.58**	2.36 **3.35**	2.28 **3.18**	2.21 **3.04**	2.15 **2.94**	2.10 **2.86**	2.06 **2.78**	2.03 **2.72**	1.89 **2.62**	1.93 **2.54**	1.87 **2.43**	1.82 **2.35**	1.78 **2.26**	1.72 **2.17**	1.69 **2.12**	1.65 **2.04**	1.62 **2.00**	1.59 **1.94**	1.56 **1.90**	1.55 **1.87**
38	4.10 **7.35**	3.25 **5.21**	2.85 **4.34**	2.62 **3.86**	2.46 **3.54**	2.35 **3.32**	2.26 **3.15**	2.19 **3.02**	2.14 **2.91**	2.09 **2.82**	2.05 **2.75**	2.02 **2.69**	1.96 **2.59**	1.92 **2.51**	1.85 **2.40**	1.80 **2.32**	1.76 **2.22**	1.71 **2.14**	1.67 **2.08**	1.63 **2.00**	1.60 **1.97**	1.57 **1.90**	1.54 **1.86**	1.53 **1.84**
40	4.08 **7.31**	3.23 **5.18**	2.84 **4.31**	2.61 **3.83**	2.45 **3.51**	2.34 **3.29**	2.25 **3.12**	2.18 **2.99**	2.12 **2.88**	2.07 **2.80**	2.04 **2.73**	2.00 **2.66**	1.95 **2.56**	1.90 **2.49**	1.84 **2.37**	1.79 **2.29**	1.74 **2.20**	1.69 **2.11**	1.66 **2.05**	1.61 **1.97**	1.59 **1.94**	1.55 **1.88**	1.53 **1.84**	1.51 **1.81**
42	4.07 **7.27**	3.22 **5.15**	2.83 **4.29**	2.59 **3.80**	2.44 **3.49**	2.32 **3.26**	2.24 **3.10**	2.17 **2.96**	2.11 **2.86**	2.06 **2.77**	2.02 **2.70**	1.99 **2.64**	1.94 **2.54**	1.89 **2.46**	1.82 **2.35**	1.78 **2.26**	1.73 **2.17**	1.68 **2.08**	1.64 **2.02**	1.60 **1.94**	1.57 **1.91**	1.54 **1.85**	1.51 **1.80**	1.49 **1.78**
44	4.06 **7.24**	3.21 **5.12**	2.82 **4.26**	2.58 **3.78**	2.43 **3.46**	2.31 **3.24**	2.23 **3.07**	2.16 **2.94**	2.10 **2.84**	2.05 **2.75**	2.01 **2.68**	1.98 **2.62**	1.92 **2.52**	1.88 **2.44**	1.81 **2.32**	1.76 **2.24**	1.72 **2.15**	1.66 **2.06**	1.63 **2.00**	1.58 **1.92**	1.56 **1.88**	1.52 **1.82**	1.50 **1.78**	1.48 **1.75**
46	4.05 **7.21**	3.20 **5.10**	2.81 **4.24**	2.57 **3.76**	2.42 **3.44**	2.30 **3.22**	2.22 **3.05**	2.14 **2.92**	2.09 **2.82**	2.04 **2.73**	2.00 **2.66**	1.97 **2.60**	1.91 **2.50**	1.87 **2.42**	1.80 **2.30**	1.75 **2.22**	1.71 **2.13**	1.65 **2.04**	1.62 **1.98**	1.57 **1.90**	1.54 **1.86**	1.51 **1.80**	1.48 **1.76**	1.46 **1.72**
48	4.04 **7.19**	3.19 **5.08**	2.80 **4.22**	2.56 **3.74**	2.41 **3.42**	2.30 **3.20**	2.21 **3.04**	2.14 **2.90**	2.08 **2.80**	2.03 **2.71**	1.99 **2.64**	1.96 **2.58**	1.90 **2.48**	1.86 **2.40**	1.79 **2.28**	1.74 **2.20**	1.70 **2.11**	1.64 **2.02**	1.61 **1.96**	1.56 **1.83**	1.53 **1.84**	1.50 **1.78**	1.47 **1.73**	1.45 **1.70**

50	4.03 **7.17**	3.18 **5.06**	2.79 **4.20**	2.56 **3.72**	2.40 **3.41**	2.29 **3.18**	2.20 **3.02**	2.13 **2.88**	2.07 **2.78**	2.02 **2.70**	1.98 **2.62**	1.95 **2.56**	1.90 **2.46**	1.85 **2.39**	1.78 **2.26**	1.74 **2.18**	1.69 **2.10**	1.63 **2.00**	1.60 **1.94**	1.55 **1.86**	1.52 **1.82**	1.48 **1.76**	1.46 **1.71**	1.44 **1.68**
55	4.02 **7.12**	3.17 **5.01**	2.78 **4.16**	2.54 **3.68**	2.38 **3.37**	2.27 **3.15**	2.18 **2.98**	2.11 **2.85**	2.05 **2.75**	2.00 **2.66**	1.97 **2.59**	1.93 **2.53**	1.88 **2.43**	1.83 **2.35**	1.76 **2.23**	1.72 **2.15**	1.67 **2.06**	1.61 **1.96**	1.58 **1.90**	1.52 **1.82**	1.50 **1.78**	1.46 **1.71**	1.43 **1.66**	1.41 **1.64**
60	4.00 **7.08**	3.15 **4.98**	2.76 **4.13**	2.52 **3.65**	2.37 **3.34**	2.25 **3.12**	2.17 **2.95**	2.10 **2.82**	2.04 **2.72**	1.99 **2.63**	1.95 **2.56**	1.92 **2.50**	1.86 **2.40**	1.81 **2.32**	1.75 **2.20**	1.70 **2.12**	1.65 **2.03**	1.59 **1.93**	1.56 **1.87**	1.50 **1.79**	1.48 **1.74**	1.44 **1.68**	1.41 **1.63**	1.39 **1.60**
65	3.99 **7.04**	3.14 **4.95**	2.75 **4.10**	2.51 **3.62**	2.36 **3.31**	2.24 **3.09**	2.15 **2.93**	2.08 **2.79**	2.02 **2.70**	1.98 **2.61**	1.94 **2.54**	1.90 **2.47**	1.85 **2.37**	1.80 **2.30**	1.73 **2.18**	1.68 **2.09**	1.63 **2.00**	1.57 **1.90**	1.54 **1.84**	1.49 **1.76**	1.46 **1.71**	1.42 **1.64**	1.39 **1.60**	1.37 **1.56**
70	3.98 **7.01**	3.13 **4.92**	2.74 **4.08**	2.50 **3.60**	2.35 **3.29**	2.32 **3.07**	2.14 **2.91**	2.07 **2.77**	2.01 **2.67**	1.97 **2.59**	1.93 **2.51**	1.89 **2.45**	1.84 **2.35**	1.79 **2.28**	1.72 **2.15**	1.67 **2.07**	1.62 **1.98**	1.56 **1.88**	1.53 **1.82**	1.47 **1.74**	1.45 **1.69**	1.40 **1.63**	1.37 **1.56**	1.35 **1.53**
80	3.96 **6.96**	3.11 **4.88**	2.72 **4.04**	2.48 **3.56**	2.33 **3.25**	2.21 **3.04**	2.12 **2.87**	2.05 **2.74**	1.99 **2.64**	1.95 **2.55**	1.91 **2.48**	1.88 **2.41**	1.82 **2.32**	1.77 **2.24**	1.70 **2.11**	1.65 **2.03**	1.60 **1.94**	1.54 **1.84**	1.51 **1.78**	1.45 **1.70**	1.42 **1.65**	1.38 **1.57**	1.35 **1.52**	1.32 **1.49**
100	3.94 **6.90**	3:09 **4.82**	2.70 **3.98**	2.46 **3.51**	2.30 **3.20**	2.19 **2.99**	2.10 **2.82**	2.03 **2.69**	1.97 **2.59**	1.92 **2.51**	1.88 **2.43**	1.85 **2.36**	1.79 **2.26**	1.75 **2.19**	1.68 **2.06**	1.63 **1.98**	1.57 **1.89**	1.51 **1.79**	1.48 **1.73**	1.42 **1.64**	1.39 **1.59**	1.34 **1.51**	1.30 **1.46**	1.28 **1.43**
125	3.92 **6.84**	3.07 **4.78**	2.68 **3.94**	2.44 **3.47**	2.29 **3.17**	2.17 **2.95**	2.08 **2.79**	2.01 **2.65**	1.95 **2.56**	1.90 **2.47**	1.86 **2.40**	1.83 **2.33**	1.77 **2.23**	1.72 **2.15**	1.65 **2.03**	1.60 **1.94**	1.55 **1.85**	1.49 **1.75**	1.45 **1.68**	1.39 **1.59**	1.36 **1.54**	1.31 **1.46**	1.27 **1.40**	1.25 **1.37**
150	3.91 **6.81**	3.06 **4.75**	2.67 **3.91**	2.43 **3.44**	2.27 **3.13**	2.16 **2.92**	2.07 **2.76**	2.00 **2.62**	1.94 **2.53**	1.89 **2.44**	1.85 **2.37**	1.82 **2.30**	1.76 **2.20**	1.71 **2.12**	1.64 **2.00**	1.59 **1.91**	1.54 **1.83**	1.47 **1.72**	1.44 **1.66**	1.37 **1.56**	1.34 **1.51**	1.29 **1.43**	1.25 **1.37**	1.22 **1.33**
200	3.89 **6.76**	3.04 **4.71**	2.65 **3.88**	2.41 **3.41**	2.26 **3.11**	2.14 **2.90**	2.05 **2.73**	1.98 **2.60**	1.92 **2.50**	1.87 **2.41**	1.83 **2.34**	1.80 **2.28**	1.74 **1.17**	1.69 **2.09**	1.62 **1.97**	1.57 **1.88**	1.52 **1.79**	1.45 **1.69**	1.42 **1.62**	1.35 **1.53**	1.32 **1.48**	1.26 **1.39**	1.22 **1.33**	1.19 **1.28**
400	3.86 **6.70**	3.02 **4.66**	2.62 **3.83**	2.39 **3.36**	2.23 **3.06**	2.12 **2.85**	2.03 **2.69**	1.96 **2.55**	1.90 **2.46**	1.85 **2.37**	1.81 **2.29**	1.78 **2.23**	1.72 **2.12**	1.67 **2.04**	1.60 **1.92**	1.54 **1.84**	1.49 **1.74**	1.42 **1.64**	1.38 **1.57**	1.32 **1.47**	1.28 **1.42**	1.22 **1.32**	1.16 **1.24**	1.13 **1.19**
1000	3.85 **6.66**	3.00 **4.62**	2.61 **3.80**	2.38 **3.34**	2.22 **3.04**	2.10 **2.82**	2.02 **2.66**	1.95 **2.53**	1.89 **2.43**	1.84 **2.34**	1.80 **2.26**	1.76 **2.20**	1.70 **2.09**	1.65 **2.01**	1.58 **1.89**	1.53 **1.81**	1.47 **1.71**	1.41 **1.61**	1.36 **1.54**	1.30 **1.44**	1.26 **1.38**	1.19 **1.28**	1.13 **1.19**	1.08 **1.11**
	3.84 **6.64**	2.99 **4.60**	2.60 **3.78**	2.37 **3.32**	2.21 **3.02**	2.09 **2.80**	2.01 **2.64**	1.94 **2.51**	1.88 **2.41**	1.83 **2.32**	1.79 **2.24**	1.75 **2.18**	1.69 **2.07**	1.64 **1.99**	1.57 **1.87**	1.52 **1.79**	1.46 **1.69**	1.40 **1.59**	1.35 **1.52**	1.28 **1.41**	1.24 **1.36**	1.17 **1.25**	1.11 **1.15**	1.00 **1.00**

Taken by permission from Paul G. Hoel, *Elementary Statistics*, Table XI. New York: Wiley, 1966.

Table 4 Durbin-Watson Critical Value Bounds
(Level of significance $\alpha = .05$)

	$q = 1$		$q = 2$		$q = 3$		$q = 4$		$q = 5$	
n	d_L	d_U	d_L	d_U	d_L	d_U	d_L	d_U	d_L	d_U
15	1.08	1.36	0.95	1.54	0.82	1.75	0.69	1.97	0.56	2.21
16	1.10	1.37	0.98	1.54	0.86	1.73	0.74	1.93	0.62	2.15
17	1.13	1.38	1.02	1.54	0.90	1.71	0.78	1.90	0.67	2.10
18	1.16	1.39	1.05	1.53	0.93	1.69	0.82	1.87	0.71	2.06
19	1.18	1.40	1.08	1.53	0.97	1.68	0.86	1.85	0.75	2.02
20	1.20	1.41	1.10	1.54	1.00	1.68	0.90	1.83	0.79	1.99
21	1.22	1.42	1.13	1.54	1.03	1.67	0.93	1.81	0.83	1.96
22	1.24	1.43	1.15	1.54	1.05	1.66	0.96	1.80	0.86	1.94
23	1.26	1.44	1.17	1.54	1.08	1.66	0.99	1.79	0.90	1.92
24	1.27	1.45	1.19	1.55	1.10	1.66	1.01	1.78	0.93	1.90
25	1.29	1.45	1.21	1.55	1.12	1.66	1.04	1.77	0.95	1.89
26	1.30	1.46	1.22	1.55	1.14	1.65	1.06	1.76	0.98	1.88
27	1.32	1.47	1.24	1.56	1.16	1.65	1.08	1.76	1.01	1.86
28	1.33	1.48	1.26	1.56	1.18	1.65	1.10	1.75	1.03	1.85
29	1.34	1.48	1.27	1.56	1.20	1.65	1.12	1.74	1.05	1.84
30	1.35	1.49	1.28	1.57	1.21	1.65	1.14	1.74	1.07	1.83
31	1.36	1.50	1.30	1.57	1.23	1.65	1.16	1.74	1.09	1.83
32	1.37	1.50	1.31	1.57	1.24	1.65	1.18	1.73	1.11	1.82
33	1.38	1.51	1.32	1.58	1.26	1.65	1.19	1.73	1.13	1.81
34	1.39	1.51	1.33	1.58	1.27	1.65	1.21	1.73	1.15	1.81
35	1.40	1.52	1.34	1.58	1.28	1.65	1.22	1.73	1.16	1.80
36	1.41	1.52	1.35	1.59	1.29	1.65	1.24	1.73	1.18	1.80
37	1.42	1.53	1.36	1.59	1.31	1.66	1.25	1.72	1.19	1.80
38	1.43	1.54	1.37	1.59	1.32	1.66	1.26	1.72	1.21	1.79
39	1.43	1.54	1.38	1.60	1.33	1.66	1.27	1.72	1.22	1.79
40	1.44	1.54	1.39	1.60	1.34	1.66	1.29	1.72	1.23	1.79
45	1.48	1.57	1.43	1.62	1.38	1.67	1.34	1.72	1.29	1.78
50	1.50	1.59	1.46	1.63	1.42	1.67	1.38	1.72	1.34	1.77
55	1.53	1.60	1.49	1.64	1.45	1.68	1.41	1.72	1.38	1.77
60	1.55	1.62	1.51	1.65	1.48	1.69	1.44	1.73	1.41	1.77
65	1.57	1.63	1.54	1.66	1.50	1.70	1.47	1.73	1.44	1.77
70	1.58	1.64	1.55	1.67	1.52	1.70	1.49	1.74	1.46	1.77
75	1.60	1.65	1.57	1.68	1.54	1.71	1.51	1.74	1.49	1.77
80	1.61	1.66	1.59	1.69	1.56	1.72	1.53	1.74	1.51	1.77
85	1.62	1.67	1.60	1.70	1.57	1.72	1.55	1.75	1.52	1.77
90	1.63	1.68	1.61	1.70	1.59	1.73	1.57	1.75	1.54	1.78
95	1.64	1.69	1.62	1.71	1.60	1.73	1.58	1.75	1.56	1.78
100	1.65	1.69	1.63	1.72	1.61	1.74	1.59	1.76	1.57	1.78

Table 4 (Continued)
(Level of significance $\alpha = .01$)

	$q = 1$		$q = 2$		$q = 3$		$q = 4$		$q = 5$	
n	d_L	d_U	d_L	d_U	d_L	d_U	d_L	d_U	d_L	d_U
15	0.81	1.07	0.70	1.25	0.59	1.46	0.49	1.70	0.39	1.96
16	0.84	1.09	0.74	1.25	0.63	1.44	0.53	1.66	0.44	1.90
17	0.87	1.10	0.77	1.25	0.67	1.43	0.57	1.63	0.48	1.85
18	0.90	1.12	0.80	1.26	0.71	1.42	0.61	1.60	0.52	1.80
19	0.93	1.13	0.83	1.26	0.74	1.41	0.65	1.58	0.56	1.77
20	0.95	1.15	0.86	1.27	0.77	1.41	0.68	1.57	0.60	1.74
21	0.97	1.16	0.89	1.27	0.80	1.41	0.72	1.55	0.63	1.71
22	1.00	1.17	0.91	1.28	0.83	1.40	0.75	1.54	0.66	1.69
23	1.02	1.19	0.94	1.29	0.86	1.40	0.77	1.53	0.70	1.67
24	1.04	1.20	0.96	1.30	0.88	1.41	0.80	1.53	0.72	1.66
25	1.05	1.21	0.98	1.30	0.90	1.41	0.83	1.52	0.75	1.65
26	1.07	1.22	1.00	1.31	0.93	1.41	0.85	1.52	0.78	1.64
27	1.09	1.23	1.02	1.32	0.95	1.41	0.88	1.51	0.81	1.63
28	1.10	1.24	1.04	1.32	0.97	1.41	0.90	1.51	0.83	1.62
29	1.12	1.25	1.05	1.33	0.99	1.42	0.92	1.51	0.85	1.61
30	1.13	1.26	1.07	1.34	1.01	1.42	0.94	1.51	0.88	1.61
31	1.15	1.27	1.08	1.34	1.02	1.42	0.96	1.51	0.90	1.60
32	1.16	1.28	1.10	1.35	1.04	1.43	0.98	1.51	0.92	1.60
33	1.17	1.29	1.11	1.36	1.05	1.43	1.00	1.51	0.94	1.59
34	1.18	1.30	1.13	1.36	1.07	1.43	1.01	1.51	0.95	1.59
35	1.19	1.31	1.14	1.37	1.08	1.44	1.03	1.51	0.97	1.59
36	1.21	1.32	1.15	1.38	1.10	1.44	1.04	1.51	0.99	1.59
37	1.22	1.32	1.16	1.38	1.11	1.45	1.06	1.51	1.00	1.59
38	1.23	1.33	1.18	1.39	1.12	1.45	1.07	1.52	1.02	1.58
39	1.24	1.34	1.19	1.39	1.14	1.45	1.09	1.52	1.03	1.58
40	1.25	1.34	1.20	1.40	1.15	1.46	1.10	1.52	1.05	1.58
45	1.29	1.38	1.24	1.42	1.20	1.48	1.16	1.53	1.11	1.58
50	1.32	1.40	1.28	1.45	1.24	1.49	1.20	1.54	1.16	1.59
55	1.36	1.43	1.32	1.47	1.28	1.51	1.25	1.55	1.21	1.59
60	1.38	1.45	1.35	1.48	1.32	1.52	1.28	1.56	1.25	1.60
65	1.41	1.47	1.38	1.50	1.35	1.53	1.31	1.57	1.28	1.61
70	1.43	1.49	1.40	1.52	1.37	1.55	1.34	1.58	1.31	1.61
75	1.45	1.50	1.42	1.53	1.39	1.56	1.37	1.59	1.34	1.62
80	1.47	1.52	1.44	1.54	1.42	1.57	1.39	1.60	1.36	1.62
85	1.48	1.53	1.46	1.55	1.43	1.58	1.41	1.60	1.39	1.63
90	1.50	1.54	1.47	1.56	1.45	1.59	1.43	1.61	1.41	1.64
95	1.51	1.55	1.49	1.57	1.47	1.60	1.45	1.62	1.42	1.64
100	1.52	1.56	1.50	1.58	1.48	1.60	1.46	1.63	1.44	1.65

By permission of authors and Biometrica Trust from J. Durbin and G. S. Watson, "Testing for Serial Correlation in Least Squares Regression," *Biometrica*, June 1951.

Index